Albert Alomba A.

The evolution of living beings

AF666782

Albert Alomba A.

The evolution of living beings

Fourth Year Scientific

ScienciaScripts

Imprint
Any brand names and product names mentioned in this book are subject to trademark, brand or patent protection and are trademarks or registered trademarks of their respective holders. The use of brand names, product names, common names, trade names, product descriptions etc. even without a particular marking in this work is in no way to be construed to mean that such names may be regarded as unrestricted in respect of trademark and brand protection legislation and could thus be used by anyone.

Cover image: www.ingimage.com

This book is a translation from the original published under ISBN 978-620-6-72424-7.

Publisher:
Sciencia Scripts
is a trademark of
Dodo Books Indian Ocean Ltd. and OmniScriptum S.R.L publishing group

120 High Road, East Finchley, London, N2 9ED, United Kingdom
Str. Armeneasca 28/1, office 1, Chisinau MD-2012, Republic of Moldova, Europe
Printed at: see last page
ISBN: 978-620-8-17663-1

Copyright © Albert Alomba A.
Copyright © 2024 Dodo Books Indian Ocean Ltd. and OmniScriptum S.R.L publishing group

Life and Earth Sciences

Albert ALOMBA A.

THE EVOLUTION OF LIVING BEINGS

Fourth Year Scientific

August 2024

FROM THE SAME AUTHOR

Cell biology for the 3ème year of scientific humanities

Biology variability approach for the students of senior six advanced level

Chemistry for the 1ère year of scientific humanities

Chemistry for the 3ème year of scientific humanities

Chemistry for the 4ème year of scientific humanities

Ecology for scientific humanities students

Evolution of living beings for the 4ème year of scientific humanities

Introduction to geology for science humanities students

Introduction to taxonomy for humanities science students

Population and natural resources for senior six advanced level

Review of plant physiology for students of senior six advanced level

SVT for the 3ème year of scientific humanities

SVT for the 4ème year of scientific humanities

SVT for grade 8 basic education

SVT for grade 7ème basic education

Biology tutorials for science humanities students

Table of contents

PREFACE

L'évolution des êtres vivants is a book for students in the fourth year of scientific humanities, preparing for state exams. in the Democratic Republic of Congo.

The notional content of the book is divided into four chapters, in the order in which they are presented: introduction to transformism, mechanisms of evolution, arguments in favor of evolution and explanatory theories of evolution.

L'évolution des êtres vivants for the fourth year of scientific humanities is written in fluent French, a simple but thorough style, so that even students for whom French is not their first language can make use of it.

All the main points covered by the official program are developed in depth, with abundant illustrations and revision questions.

It's up to the teacher in charge of the life and earth sciences course in the fourth year of the Scientific Humanities to be ingenious in selecting the topics he or she will be able to develop for students in the scientific section, and those he or she will not be able to develop for students in other sections, such as the pedagogical or literary sections.

We would like to extend our anticipatory thanks to fellow life and earth science teachers who, through positive criticism, will be willing to improve future editions of this book.

Contact details, e-mails and telephone numbers are given on the last page of this book.

Albert ALOMBA

CHAPTER 1: INTRODUCTION TO TRANSFORMISM

1.1 FIXISM OR TRANSFORMISM

1.1.1 FIXISM

Until the 18th century, scientists believed that the earth had always been populated by the same flora and fauna as it is today. This was the fixist doctrine.

So, to explain the origin of living species, some called on **Creationism**, others on **Spontaneous Generation**.

1.1.1.1 CREATIONISM

The doctrine that **God created everything**. In the Bible, for example, we read this in Genesis chapter 1, verse 1: In the beginning, God created the heaven and the earth. Or in Revelation chapter 4, verse 11: You are worthy, Jehovah our God, to receive glory and honor and power, for you created all things, and because of your will they came into being and were created, end of quote.

1.1.1.2 SPONTANEOUS GENERATION

This refers to **the spontaneous**, sudden and unforeseeable **appearance** of living beings from non-living matter under favorable environmental conditions.

Here are some defenders of fixism:

GEOFFROY-SAINT HILAIRE: (French, 1772-1844) created embryology, laying the foundations for comparative anatomy.

Etienne Geoffroy Saint-Hilaire
(1772-1844)

GEORGES CUVIER: (French 1769-1832), created paleontology, the science of studying fossils. He established the principle of organ correlation.

Here's his **hypothesis**: the shape of one organ leads to that of the others, so if we only know one part of an animal, it should enable us to find the organization of the other parts of the same animal. This proves true, and fossil parts and fragments can also be used.

Georges Cuvier
(1769-1832)

1.1.2 TRANSFORMISM

Around the beginning of the 19ème century, the idea of **transformism** began to emerge, or evolution, according to which present-day living forms derive from ancient, simpler forms. For evolutionists, the idea of transformism is the only scientific explanation for the fundamental characteristics of the world today and in the past.

The Frenchman **Lamarck**, the Englishman **Darwin** and the Dutchman **Hugo De Vries** are among the proponents of the doctrine of evolution.

Other theories were born in an attempt to find the origin of the living world:

The **synthetic theory of evolution is** a kind of synthesis between Lamarckism, Darwinism and mutationism, while rejecting Lamarck's heredity of acquired traits.

Cell theory, i.e. the notion that the living cell is the fundamental unit of all living animal and plant organisms, was born in 1839 from the confrontation of the work of M.J. SCHLEIDEN (Botanist :1804-1881) and T. SCHWAN (Zoologist :1810-1880).

Matthias Jacob Schleiden
(1804-1881)

Theodor Schwann
(1810-1880)

Cell theory bridges the gap between the most dissimilar living beings. This theory was completed by Rudolf VIRCHOW (1821-1902) with his **postulate** "*Omni cellula e cellula*".

Rudolf VIRCHOW
(1821-1902)

The **Big Bang theory**: according to the theory of general relativity, an expanding Universe must have begun with a Big Bang, implying the finiteness of **time**, which appeared simultaneously with **space** and **energy-matter**. This view has changed in recent years.

Figure 1.1 *The big bang*

The Catholic priest Abbé **Georges Lemaitre** (Belgian astronomer and physicist) is the author of the **Big Bang** theory. In 1927, he was the first to propose that the universe began with an explosion.

Georges Lemaitre
(1894-1966)

Edwin Hubble (American astronomer and physicist), found experimental evidence to support Lemaître's theory. He discovered that distant galaxies are moving away from us in all directions, with speeds proportional to their distance.

Edwin Hubble
(1889-1953)

Three concepts emerge from this presentation of **evolution**:

- Life appeared on earth millions of years ago, through the interplay of physico-chemical forces acting according to the laws of chance.
- This life first materialized in rudimentary form, then became progressively more complex over the ages, always according to the laws of chance. As a result, living forms have changed. Species gave birth to other, more organized species, different from the first.
- The idea of progress is slow: life rises from the inorganic to the organic, from the simple to the complex, from primitive gelatin to man, through the most diverse forms.

1.2 ORIGIN OF A THEORY

The theory of evolution, evolution or change over time, is the process by which the modern organism descended from an ancient, primitive organism.

The groundwork for modern evolutionary theory was laid in the 1700s and 1800s.

1785: **James Hutton**. Hutton proposes that the Earth is shaped by **geological forces** that took place over extremely long periods of time. He believes that the Earth is millions, not thousands, of years old.

James Hutton
(1726 à 1797)

1798: Thomas Malthus

In his essay on the *population principle,* Malthus predicts that human populations will grow faster than space and food resources can support them.

Thomas Malthus
(1766-1834)

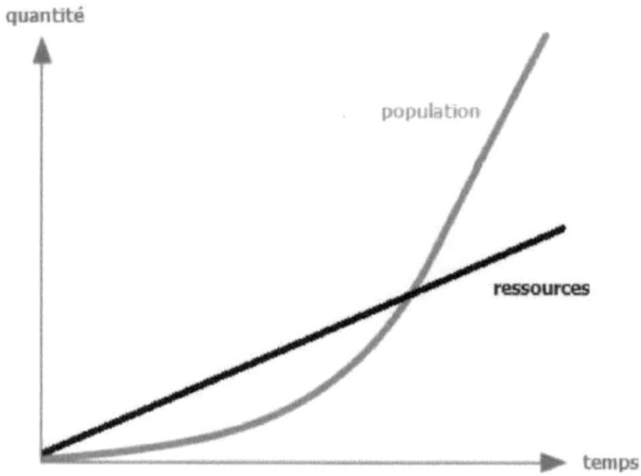

Figure 1.2 *Malthus chart*

1809: Jean-Baptiste Lamarck

Jean-Baptiste de Lamarck
(1744-1829)

In *Zoological Philosophy*, Lamarck published his hypotheses on the heredity of acquired traits. The ideas were wrong, but he was one of the first to propose a mechanism to explain how organisms change over time.

1831: Charles Darwin

Charles Darwin
(1809-1882)

Darwin embarked on the H.M.S. Beagle, a voyage that would provide him with vast quantities of evidence leading to the theory of evolution. Darwin published his ideas in *The Origin of Species* (1859).

NB HMS is a Royal Navy ship, Her Majesty's Ship or His Majesty's Ship.

Figure 1.3 *HMS Beagle*

1833: Charles Lyell

In the second and final volume of *Principle of Geology,* **Lyell** explains that the processes occurring now have shaped the Earth's geological features over a long period of time.

Charles Lyell
(1767-1849)

1858: Alfred Wallace

Alfred Russel Wallace
(1823 -1913)

Wallace wrote to Darwin, speculating on evolution by natural selection based on his studies of the distribution of plants and animals. Darwin and Wallace jointly presented their *theory of evolution and natural selection* to the Linnaean Society in 1858.

1910: Hugo De Vries

Hugo De Vries
(1848-1935)

In *Die mutations,* De Vries presents his evolutionary theory of mutations.

1.3 THE ORIGIN OF LIFE 1.3.1 BIOGENESIS

The principle of **biogenesis**, which states that all living things come from other living things, seems very reasonable today. Before the 17th century, however, it was generally thought that living things could also arise from the non-living environment, through a process called **spontaneous generation,** if that environment offered favorable conditions.

This seemed to explain why maggots appeared on rotting meat and why fish appeared in ponds that had been drained the previous season, people thought the mud might have given birth to the fish.

Scientists carried out controlled experiments to try and better understand the process of spontaneous generation.

REDI EXPERIMENTS

The Italian scientist Francisco Redi noticed and described the different forms of fly development.

Francesco Redi
(1626-1697)

In Redi's experiment, maggots were only found in the control pots, as this was the only place where adult flies could reach the meat to lay their eggs.

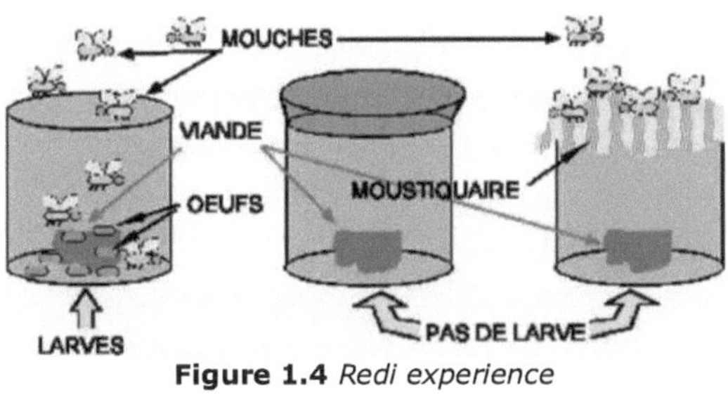

Figure 1.4 *Redi experience*

JOHN NEEDHAM'S EXPERIENCE

To support the thesis of spontaneous generation, Needham published the results of his experiments. These consisted in pouring a meat broth into a bowl, sealing the bowl hermetically - according to him - and then reheating it by immersing it for a few moments in hot ashes.

John Needham
(1731-1781)

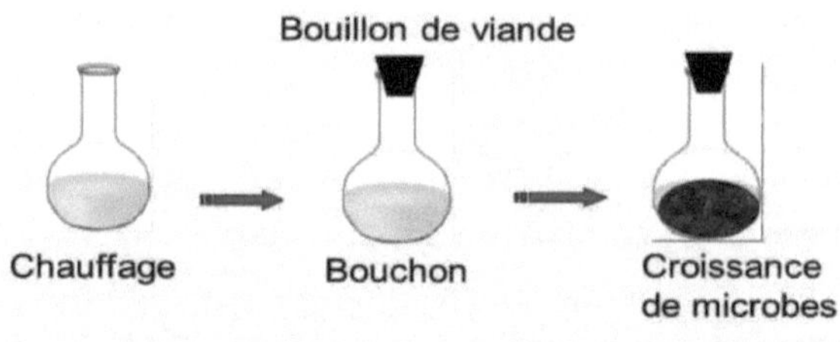

Figure 1.5 *Needham's erroneous experience*

THE SPALLANZANI EXPERIENCE

In the 1700s, another Italian scientist, Lazzaro Spallanzani, devised an experiment to test the hypothesis of spontaneous generation.

Lazzaro S pa l l anzani (1729-1799)

In Spallanzani's experiment, he boiled meat broth in open vials and then sealed the vials of the experimental group by melting the glass necks of the closed vials. The broth inside remained uncontaminated by micro-organisms.

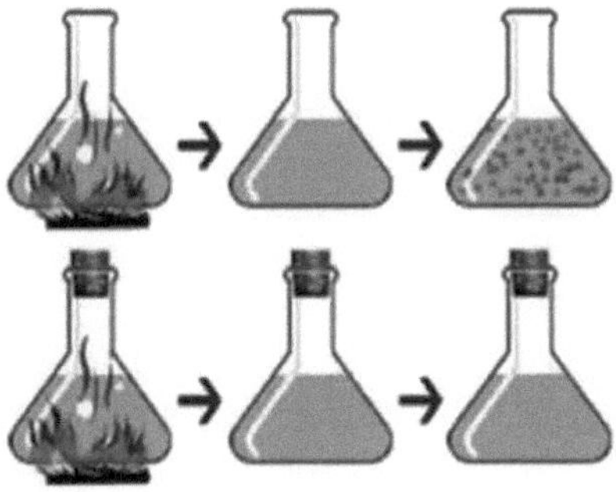

Figure 1.6 *Spallanzani experiments*

THE PASTEUR EXPERIENCE

By the mid-1800s, the controversy over spontaneous generation had become fierce. The Paris Academy of Sciences offered a prize to anyone who could clear up the problem once and for all. The prizewinner was French scientist **Louis Pasteur** (1822-1895).

In Pasteur's experiment, a flask with a curved but open neck prevented micro-organisms from entering the broth boiled in the flasks was only contaminated with micro-organisms when the curved necks were removed from the flasks.

Louis Pasteur
(1822-1895)

Louis Pasteur put an end to the abiogenesis debate with his swan-neck flask experiment. He is the father of **microbiology**.

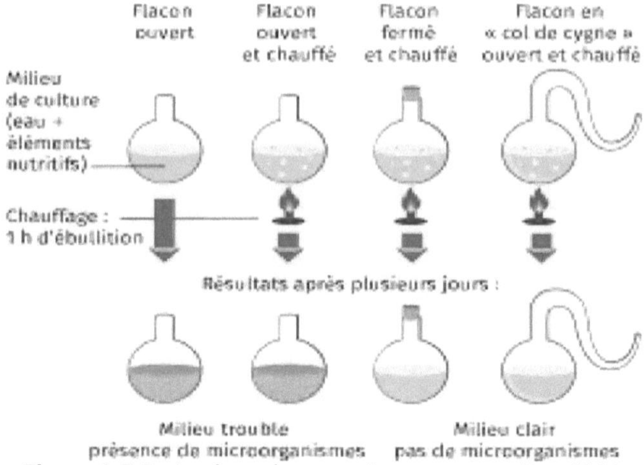

Figure 1.7 *Pasteur's work on spontaneous generation, 1864.*

1.3.2 ORIGIN OF WATER ON EARTH

The **origin of water** on earth has not yet been elucidated. There are also competing theories as to how water came to be on earth. One says that comets, some of which contain ice, sank into the earth and provided all our water.

It's true that the primitive earth was regularly bombarded by **asteroids**, **comets** and **meteors**. Whether this resulted in the earth's oceans is less certain. The competing school of thought believes that water was already present in the earth when it formed. Given that water is not entirely scarce in space, it's possible that this is also correct.

Water on the primitive earth was a little different from what we have today. The first oceans were, like today, salty. Scientists believe that the first oceans were about 1.2 to 2 times saltier than our own. The salt in the seas came mainly from volcanic activity and underwater volcanic rock. This volcanic activity also contributed to the creation of fresh water. It is likely that volcanic island chains were brought together by the movement of tectonic plates, forming the core of the first continents. Once there was land, freshwater accumulated.

The formation of the earth's oceans was eventually followed by the first life on earth. The first forms of life on earth could exist thanks to water.

1.3.3 THE FIRST ORGANIC COMPOUNDS

All the elements found in organic compounds are thought to have existed on Earth and in the rest of the solar system when the Earth formed.

Alexander Oparin
(1894-1980)

In 1923, Soviet scientist **Alexander Oparin** (1894-1980) suspected that the early Earth's atmosphere was very different from that of today. Oparin believed that the primitive atmosphere contained ammonia NH_3, hydrogen gas H_2, water vapor H_2O and compounds of hydrogen and carbon, such as methane CH_4.

At temperatures well above the boiling point of water, these gases could have formed simple organic compounds, such as amino acids.

According to **Oparin**, when the Earth cooled and water vapor condensed to form lakes and the sea, these simple organic compounds would have accumulated in the water.

Over time, these compounds may have entered complex chemical reactions, fueled by the energy of lightning and ultraviolet radiation. These reactions, according to Oparin, would eventually have led to the macromolecules essential to life, such as proteins.

1.3.4 EXPERIMENTAL SYNTHESIS OF ORGANIC COMPOUNDS: THE MILLER AND UREY EXPERIMENT

Oparin carefully developed his hypotheses, but didn't carry out any experiments to test them. So, in 1953, American graduate student **Stanley Miller** (1930-), and his professor **Harold Urey** (1893-1981), set up an experiment using Oparin's hypotheses as a starting point.

Stanley Miller
(1930-2007)

The **Miller-Urey** experiment and subsequent variations produced a variety of organic compounds, including amino acids.

Since the 1950s, scientists have continued to explore the origins of simple organic compounds, including various amino acids, ATP and DNA nucleotides. Such findings suggest numerous ways in which vital organic compounds could have formed on the young Earth.

Harold Urey
(1893-1981)

Sidney Fox (1912-) and others have done extensive research into the physical structures that may have given rise to the first cells.

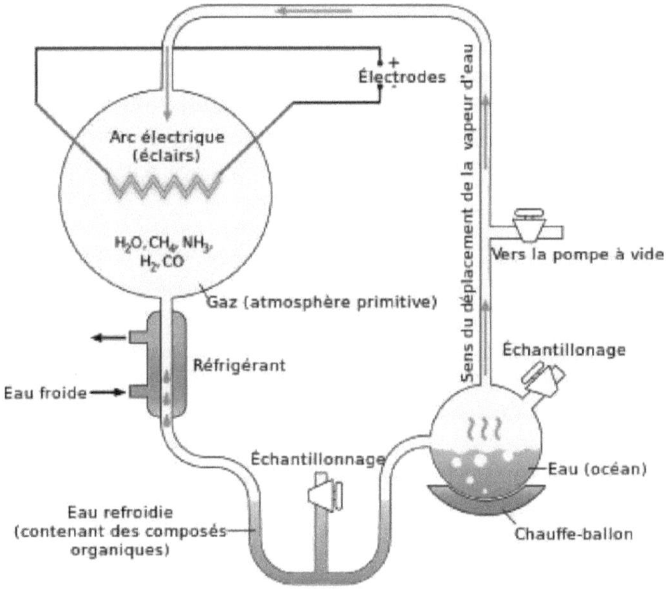

Figure 1.8 *Miller-Urey experiment*

1.3.5 THE FIRST PROKARYOTES

When the **first organisms** appeared, there was little or no oxygen gas. So the first cells must have been anaerobic. The small size of the oldest microfossil indicates that these first cells were **prokaryotes**. These cells were probably heterotrophs, absorbing organic molecules from their environment.

The first autotrophs, however, probably didn't depend on photosynthesis as most autotrophs do today. If we look at living organisms that may be similar to these early organisms, we find **archaea**.

Archaea are a kingdom of single-celled organisms, many of which thrive in extremely harsh environmental conditions. *Methanosarsina barkeri, for example,* lives in anaerobic marine sediments. Many species of archaeobacteria are autotrophs, obtaining energy by **chemosynthesis** instead of photosynthesis. In the process of chemosynthesis, CO_2 serves as the carbon source for the assembly of organic molecules. Energy is obtained from the oxidation of various inorganic substances, such as sulfur.

Photosynthesis and aerobic respiration, oxygen, a by-product of photosynthesis, damaged many early single-celled organisms. Oxygen could destroy certain coenzymes essential to cell function in some organisms, however, **oxygen binds** to other compounds, preventing oxygen from doing any damage. This **binding** was one of the first steps in aerobic respiration.

Thus, an early function of aerobic respiration may have been to prevent the destruction of essential organic compounds by oxygen.

Some life forms had become photosynthetic 3 billion years ago In the mid-1990s, scientists discovered traces of carbon in geological formations off the coast of Greenland.

The 3.5 billion-year-old microfossils found in Australia are probably photosynthetic single-celled organisms related to **modern cyanobacteria**, a group of photosynthetic single-celled prokaryotes.

It took a billion years or more for oxygen gas levels to reach today's levels. The oxygen, O2, eventually reached the upper atmosphere, where it was bombarded by sunlight. Sunlight can split O2 to form highly reactive single oxygen atoms O.

These react with O2 to form Ozone, o3.

Ozone **is toxic to** both plant and animal life, but in the upper atmosphere, a layer of ozone absorbs the sun's intense ultraviolet radiation. UV rays damage DNA, and without the protection of the ozone layer, life could not exist on earth.

1.3.6 THE FIRST EUKARYOTES

Eukaryotic cells differ from prokaryotic cells in several ways. Eukaryotic cells are larger. Their DNA is organized into chromosomes in a cell nucleus, and they contain membrane-enclosed organelles (nucleus, mitochondria, chloroplast, ER, vacuole, Golgi apparatus...).

A large body of evidence now suggests that around 20 to 15 billion years ago, a type of small aerobic prokaryote entered and began living and reproducing in larger aerobic prokaryotes (**primitive phagocyte**).

Lynn Margulis
(1938-2011)

Researcher **Lynn Margulis** (1938-2011) suggested that what might have started as an invasion became a fruitful and mutually beneficial relationship, called endosymbiosis. It is thought that the aerobic prokaryote eventually gave rise to the modern **mitochondria**, which are the site of aerobic respiration in eukaryotic cells.

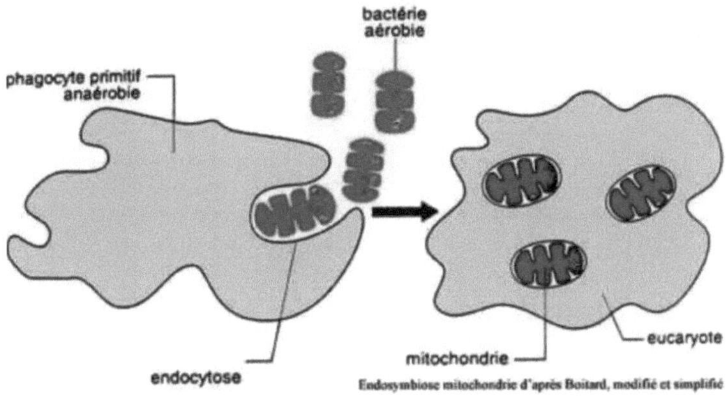

Figure 1.9 *Endosymbiotic theory, origin of mitochondria.*

Some time later, there was a second successful invasion of eukaryotic cells. This time, the invader was a relative of modern photosynthetic cyanobacteria. These invaders eventually gave rise to **chloroplasts**, the sites of photosynthesis.

There is convincing evidence to support this hypothesis **of eukaryotic evolution**, both of chloroplasts and of the replication cycle of the cell containing them.

What's more, the chloroplasts and mitochondria contain some of their own origins, which are different from those of the rest of the cell. These genes are found in the organelles themselves, in a circular piece of DNA, an arrangement characteristic of hard, non-eukaryotic prokaryotic DNA.

This process, at the origin of the mitochondrion and chloroplast, is called **symbiotic adoption**. But having lost a number of important genes, the two organelles can no longer leave the host cell and live independently of it.

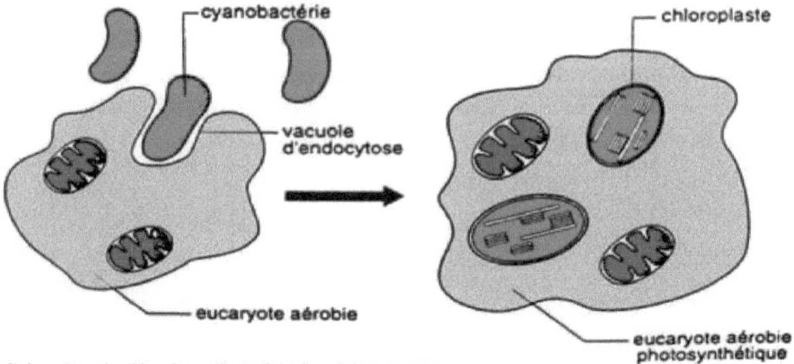

Endosymbiose des chloroplastes d'après Boitard, modifié et simplifié

Figure 1.10 *Endosymbiotic theory, origin of chloroplasts.*

REVISION QUESTIONS

1. Explain the following concepts: fixism, creationism, spontaneous generation, the big bang, biogenesis, endosymbiotic theory.
2. Name a few defenders of fixism.
3. Name a few proponents of transformism.
4. In the context of transformism, give the contribution of each of the following scientists: Georges Lemaire, Edwin Hubble, James Hutton, Charles Lyell, Thomas Malthus, Jean-Baptiste de Lamarck, Charles Darwin, Alfred Russel Wallace, Hugo De Vries.
5. Describe the experiments of the following scientists in the fight against spontaneous generation: John Needham, Francesco Redi, Lazzaro Spallanzani, Louis Pasteur.
6. Summarize Alexander Oparin's ideas and describe Stanley Miller's experiment.
7. How can you explain the endosymbiotic theory?
   (Lynn Marguillis theory, chloroplast case)

CHAPTER 2: EVOLUTIONARY MECHANISMS

2.1 THE VARIATION

Variation is the difference in characteristics between individuals of the same species or natural population.

The variation shown by members of a species may be due to :

- Differences in genetic make-up
- Environmental conditions
- Interaction effects between genetic and environmental influences.

2.1.1 SIZE OF VARIATION

- Variation is an important source of raw material for evolution.

- A population with a similar phenotype can survive in a stable, optimal environment, but will die out if the environment changes.

- Variation increases a species' ability to survive in a changing environment. There is an increased probability that certain individuals in a population will be able to tolerate a particular environmental change. Such individuals can survive and reproduce to pass on their people to their offspring.

- A population with different phenotypes may be able to inhabit a wider range of habitats and niches.

- Natural selection will increase the frequency of advantageous ones over successive generations. Over a long period of time, accumulated genetic differences can lead to speciation (the development of one or more new species from an existing one).

- A low-diversity population has a reduced capacity to adapt to environmental change. This can lead to the extinction of the population and possibly the species as a whole.

2.1.2 SOURCES OF VARIATION

Variation can come from either genetic sources or environmental factors.

2.1.2.1 GENETIC SOURCES

The main genetic sources of variation are :

- Random/independent assortment of homologous chromosomes and therefore alleles that occurs during metaphase, meiosis.

- Crossover between non-sister chromatids of homologous chromosomes during meiosis produces new allele combinations.

- Random interaction between organisms within a species during sexual reproduction creates new combinations, new genotypes of genes, and therefore new phenotypes (variants) in the population.

- The random fusion of male and female gametes during fertilization is also a source of variation.

- Chromosomal mutation.

- Genetic mutation.

Mutation is the most important source of variation, producing new alleles which can then be combined in different ways with existing genotypes.

2.1.2.2 ENVIRONMENTAL FACTORS

The main environmental factors and influences on variation are :

- In many cases, the phenotype variation caused by the combination of alleles is also influenced by the environment.

- Examples of environmental factors that can affect the phenotype of individuals in a population are: climate such as light, temperature and humidity; food/nutrient type; water; edaphic factors.

- Phenotypic variance due to environmental variation cannot be inherited.

- However, environmental selection can lead to changes in gene allele frequencies within a population. For example, a change in the environment leading to a decrease in environmental temperature will favor the survival and reproduction of a population of bears with a thicker fur layer and an increased ability to maintain their body temperature.

2.1.3 TYPES OF VARIATION

In a natural population, there are two main types of variation: continuous and discontinuous.

2.1.3.1 CONTINUOUS VARIATION OR QUANTITATIVE INHERITANCE

- There is no clearly defined difference in characteristics.

- Characteristics vary continuously from one extreme to the other, with the majority of individuals having a phenotype intermediate between the extremes.

- These traits are controlled by the additive effect of genes, or polygenics, and are called polygenic traits.

- Examples are height, body weight and skin color in humans.

2.1.3.2 DISCONTINUOUS VARIATION OR QUALITATIVE VARIATION

Discontinuous variation occurs when a characteristic is either present or absent; there are no intermediate forms.

- Discontinuous variation clearly defines the differences in certain characteristics that can be observed in a population.

- It does not show a normal distribution curve, but is classified into two or more distinct groups with no intermediate forms.

- Discontinuous variation generally determined by different alleles of the same gene.

- Phenotypic expression is generally unaffected by environmental conditions.

- Examples of human characteristics that show discontinuous variation are: Human ABO blood group, tongue roll (can or cannot), basic forms of human fingerprints, ability to taste bitter chemicals phenylthiocarbamide (PTC), taster or non-taster, ...

Table 2.1 *Comparison between continuous and discontinuous variation*

	CONTINUOUS VARIABLE	DISCONTINUOUS VARIATION
Features	- Not categories separate - No limit between values - Tendency to be quantitative	- Separate categories - Not categories intermediaries - Tendency to be qualitative
Examples	- Size - Weight - Heart rate - Finger length - Sheet length	- Human ABO blood types - Tongue roll (may or may not) - Basic human fingerprint shapes - Ability to taste products bitter chemicals phenylthiocarbamide (PTC), taster or non-taster,
Graphics	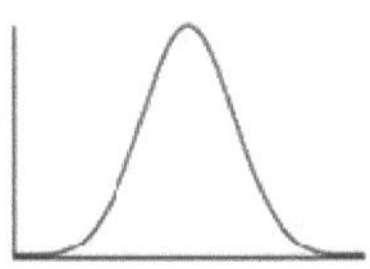 Line graph	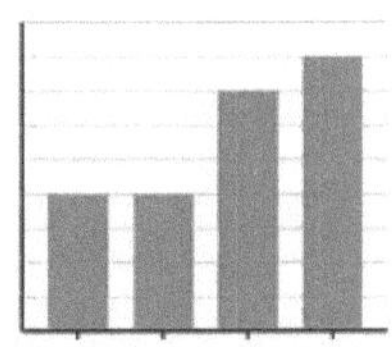 Bar graph
Control	The characteristics are controlled by the additive effect of several genes and are called characteristics. polygenic.	Discontinuous variation generally determined by different alleles of the same gene.

2.2 SELECTION

2.2.1 DEFINITION OF SELECTION

Selection is the process by which one or more factors acting on the different phenotypes within a population produce differential mortality and favor the transmission of alleles to subsequent generations.

These can confer advantages on offspring when expressed as phenotypes.

There are two types of selection: natural and artificial.

2.2.2 NATURAL SELECTION

The agent of natural selection is the environment.

Living organisms have enormous reproductive potential to produce more offspring that can eventually survive.

However, the natural population tends to remain more or less constant, as environmental factors such as available food, space or light can limit the size of a population.

There is a constant struggle for existence; the struggle for life (intraspecific and interspecific competition).

There is variation in genotypes and therefore phenotypes among individuals in a population.

The variants best adapted to their natural environment, such as abiotic conditions, predation and disease resistance, will be selected to survive and reproduce. In this way, survival of the fittest occurs.

Advantageous characteristics are heritable and will be passed on to their offspring. Selection pressures increase the chances of transmitting certain advantageous alleles to subsequent generations.

Organisms with less advantageous variants will be at a selective disadvantage. Their chances of reproducing and transmitting alleles to their offspring are diminished.

Such changes in allele frequency within a population are the basis of evolution. Over several generations, populations can

gradually evolve and become better adapted to their environmental conditions.

The main types of natural selection are: stabilizing selection, directional selection, disruptive selection, sexual selection and polymorphism.

2.2.2.1 THE DIRECTIONAL SECTION

In **directional selection**, as environmental conditions change, individuals in a population with one of the extreme phenotypes gain an advantage over the others and are selected. The range of phenotypes changes as some individuals fail to survive and reproduce, while others succeed.

Directional selection pressure increases the chances that advantageous alleles will be passed on to the next generation.

Selected pressure reduces the chances of transmission of disadvantageous alleles.

Examples:

- Industrial melanism, the increasing melanin pigmentation in the wings and body of the peppered butterfly, helps the butterfly camouflage itself when resting on a dark tree trunk near soot-polluted industrial areas to avoid visual predation by moth-eating birds.

- Increasing antibiotic resistance in pathogenic bacteria through the selection of resistant mutants.

- Increase in insecticide-resistant pests...

2.2.2.2 DISRUPTIVE SELECTION

When individuals at the upper and lower ends of the curve are in better physical condition than those near the middle, **disruptive selection** takes place.

These individuals have better survival and reproduction rates. Disruptive sectioning pressure increases the chances that advantageous alleles will be passed on to the next generation. The intermediate phenotype is selected against and progressively decreases in number from generation to generation, eventually disappearing.

After many generations, disruptive selection can lead to the formation of two distinct gene pools and the new species.

2.2.2.3 SEXUAL SELECTION

Sexual selection is the struggle generally between male animals, selected by females of the same species in the mating process. This occurs in a sexually reproducing animal population.

Sexual dimorphism is the difference in secondary sexual characteristics between male and female animals of the same species.

The success of selection by females depends on certain secondary male sexual characteristics that have evolved.

Examples: Male birds with the best courtship display, such as song or dance, birds with the brightest colored plumage, for example...

Some mammals and insects release pheromone molecules that act as chemical stimuli to attract their partners.

The alleles that determine the secondary sexual characteristics that enhance mating success are generally inherited by the male offspring.

Secondary **sexual characteristics** that give an individual an advantage in sexual selection can sometimes be neutral or disadvantageous for survival. Larger size and more colorful plumage can attract the attention of predators.

2.2.2.4 POLYMORPHISM

Polymorphism is the existence of two or more distinctly different forms or morphs of a particular species represented in a population.

In social insects such as honeybees, the caste system of queens, drones and workers is an example of **environmental polymorphism**. The differences are caused by environmental factors. The caste system of honeybees is determined by the different types of larval food supply.

There are also two types of genetic polymorphism, where differences are due to inherited characteristics. The two existing forms are: balanced polymorphism and transient polymorphism.

- Balanced polymorphism

Balanced polymorphism is the existence of two or more different distinct forms (morphs) that coexist in a stable ratio within a population. Each form may have both advantageous and disadvantageous characteristics.

Examples are :

ABO blood types in humans: There are 4 forms: type A, type B, type AB, type O.

Sickle cell disease (SCD) in Africa.

- **Transient polymorphism**

Transient polymorphism occurs when a particular form is propagating through a population, causing the alteration of a relative proportion of phenotypes. The frequency of each phenotypic form is determined by the intensity of selective pressure.

Example: Dark and light forms of pepper moth, where the light form is gradually replaced by the phenotype portion.

2.2.3 ARTIFICIAL SELECTION

2.2.3.1 DEFINITIONS

Artificial selection is selective breeding that occurs when humans, instead of environmental forces, select and determine the desirable alleles of plants or animals to pass on to successive generations.

Artificial selection has been practised by man for several centuries. It has played an important role in the evolution of modern crop plants, farm animals and domestic pets from their wild ancestors.

It exerts a directional selection pressure that leads to changes in allele and genotype frequencies in the population.

Some of the advantages of artificial selection are :

- This is the fastest and safest way to produce offspring with the desired characteristic.
- Select and breed animals and plants capable of adapting to live in certain habitats or under different environmental conditions (ecological tolerance).
- Produce pest-, disease- and herbicide-resistant organisms.
- Select and multiply crops such as wheat, barley, rice and corn for higher productivity (higher yield per unit area).
- Select and breed farm animals for better quality and quantity of milk and meat production. Select pedigree sheep to improve wool quality.
- To select and breed animals for sport or leisure. For example: breeding horses for racing and transport, breeding pigeons for flying ability and plumage type, domesticating dogs as guard dogs, for hunting, racing and appearance, breeding **koi** fish for appearance, breeding orchids, roses and other flowers to produce larger and more colorful.

The establishment of a sperm bank and a gene bank, and improvements in reproductive techniques such as artificial insemination and embryo transplants, have increased the chances of successful selective breeding in animals.

Plant and animal breeders are also using mutagens, new biotechnology and genetic engineering methods to create new phenotypes. One example is protoplast fusion, a new technique used in the production of a virus-resistant hybrid tobacco plant.

Excessive selective breeding between closely related organisms increases homozygosity (production of individuals with harmful or undesirable phenotypic characteristics). After several generations of inbreeding, vigor is lost.

Artificial selection modifies allele and genotype frequencies in a natural population. Humans have artificially modified the genome that has been selected by nature and that is perhaps best suited to the organism in its natural environment. Characteristics that appeal to humans could prove harmful to the organism.

However, some scientists argue that artificial selection and biotechnology can combine characteristics in a short space of time that natural selection would require thousands or millions of years.

2.2.3.2 HOMOGAMY AND CONSANGUINITY

This is a selective cross between individuals that have a similar genotype (homogamy) or are more closely related (inbreeding) than if they had been chosen at random from the population as a whole.

Examples of inbreeding include self-fertilization in plant mating between offspring with a parent, between siblings or closely related individuals.

After several generations, the selection force of inbreeding increases the frequency of homozygous genotypes. The organism is probably purebred or homozygous for the selected characteristics.

Inbreeding tends to maintain desirable characteristics such as :

- Increase the quantity and quality of milk produced by Jersey cows (high cream content).
- Produce corn plants of uniform height to facilitate mechanical harvesting.
- Increase oil content to reduce production and extraction costs.
- Increase crop and livestock yields. Large quantities of food can be produced, production costs fall and less land is needed for farming or livestock rearing.
- Breeding racehorses.
- Inbreeding dogs to produce competition or guard varieties.

Disadvantages include:

- After several generations of excessive inbreeding, the result is inbreeding dispersion. Inbred offspring have reduced vigor, poor growth and yield, and lower fertility than non-inbred individuals.

- Increased homozygosity can lead to individuals with undesirable homozygous recessive alleles. This would lead to the production of detrimental phenotypic characteristics.

- There is an increased risk of reduced disease resistance as genetic variation is reduced. So inbreeding is not encouraged by breeders.

2.2.3.3 HYBRIDATION

- Hybridization is the controlled mating (or cross-breeding) of distantly related individuals. They may come from two breeds of the same species, or from different species.

- It is used by plant and animal breeders to combine the desirable characteristics of two different parents.

- The offspring are known as hybrids. They generally show more variation than offspring produced by inbreeding. Hybrids generally have new, superior phenotypes and a greater capacity to adapt to environmental change.

- Hybridization increases heterozygosity and provides a new opportunity for gene interaction. Harmful recessive alleles are masked by dominant alleles.

- In some cases, inbreeding results in hybrid vigor, healthier and larger offspring.

- Hybrids produced between genetically different species are often sterile. They have no homologous chromosome sets, and meiosis cannot produce fertile gametes. However, if chromosome doubling occurs in a hybrid, then a fertile allotetraploid (polyploidy) with two sets of homologous chromosomes is produced.

- In Malaysia, the hybrid oil palm *Elaeis guineensis* variety **tenera**, with a high content and not falling easily, is produced by crossing the mother plant *E. guineensis* variety **dura** with *E. guineensis* variety **pissifera**.

2.2.3.4 HUMAN SPEMATOZOID BANK

Human sperm bank

This resource center collects, stores and freezes human sperm for future use.

Human sperm can be stored frozen for several years in tubes placed in a liquid nitrogen tank.

Human sperm banking is used in many different situations. For example, sperm is stored if :

- The person undergoes treatment for infertility. Sperm is collected and stored. Normal sperm is used later in artificial insemination or other advanced reproduction.

- The person is diagnosed with cancer and undergoes chemotherapy and radiotherapy, which can affect sperm production or cause mutation of sperm DNA.

- The person is taking certain medications, has a medical condition such as multiple sclerosis or an operation that will affect the testicles and the ability to produce sperm.

- The person will undergo a vasectomy.

- The person works in an area or laboratory and may be exposed to radiation or reproductive toxins.

Sperm can be stored frozen for several years and destroyed when instructions are given.

2.2.3.5 ANIMAL SPERM BANK

The animal sperm bank is a resource center that collects, stores and freezes animal sperm in liquid nitrogen.

Sperm can be stored for future use such as :

- Use of semen from pedigree animals with desirable characteristics for artificial insemination or in vitro insemination.

- Sperm from endangered species, such as the **Panda**, can be frozen and distributed to centers around the world for scientific research, or used in artificial insemination to fertilize the egg of the endangered female animal.

A semen sample contains thousands of sperm cells. Sperm can be diluted into smaller samples, and a small quantity is used each time.

Health regulations require that semen collected from animals be pathogen-free.

The animal semen bank also acts as a gene bank where animal semen DNA can be stored before purebred farm animals die out or endangered species become extinct.

NOTES :

- In humans, certain hereditary diseases tend to occur with greater frequency in families with a history of marriage between close relatives.

- Hybrid vigor is the superiority of one heterozygote over another homozygote for a particular trait.

- In vivo, describing biological processes as observed in the body of a living organism.

- In vitro, describing biological processes that take place outside the living organism; occurring in an artificial environment, e.g. in a test tube.

- **Cryopreservation** is the process of freezing tissue. In sperm cryopreservation, a sample of human sperm is frozen for future use.

A sperm cryobank is a center that collects, stores and freezes sperm.

2.3 THE SPECIATION

2.3.1 DEFINITION OF SPECIATION

Speciation is the process of forming one or more new species from existing ones.

In intraspeciation, a single species can give rise to several species.

In interspecific hybridization, two different species can give rise to a new species. This is common in flowering plants.

New species may be able to adapt to their particular environment, resulting in higher survival rates and more successful reproduction.

2.3.2 BIOLOGICAL SPECIES CONCEPT

There are other ways of defining a species according to different aspects or criteria. Examples are :

Reproduction: A species is a group of organisms that can reproduce and produce fertile offspring, and that are reproductively isolated from other species.

Genetic: A group of organisms with a similar genetic karyotype.

Morphology: A group of organisms with similar morphological characteristics.

Ecology: Group of organisms occupying the same ecological niche.

Species recognition: based on **behavior**.

Over the last few decades, the most common definition has been the biological species concept (BSC), but this concept has undergone a number of changes over the years.

The biological species concept (BSC) defines species as groups of actual or potentially reproducing natural populations with a common gene pool, which are **reproductively isolated** from other such groups. Organisms are therefore considered to belong to the same species if they can interbreed and produce fertile offspring. Otherwise, the organisms are considered to belong to different species.

Although CEB is quite widely accepted, there are also limitations or problems encountered:

- Some populations reproduce asexually (these include bacteria, protozoa, euglenoid flagellates and some plants).

- Groups of organisms that do not exist together in time cannot be evaluated, it is difficult to classify extinct fossilized organisms, ...

A phylogenetic species concept has been proposed in which the grouping of related species is based on shared DNA/DNA homology and the evolutionary history of the population, rather than on interbreeding between organisms.

2.3.3 SPECIATION PROCESS

The formation of new species can be caused by: isolation, genetic drift, hybridization and adaptive radiation.

2.3.3.1 ISOLATION

Isolation divides a large population physically or reproductively into smaller populations. Isolation is important in speciation. There are two types of isolation: extrinsic and intrinsic.

Extrinsic or geographical isolation

- Extrinsic isolation is also known as geographical isolation. It is caused by geographical barriers such as mountains, oceans, deserts, rivers and glaciers.

- Geographical isolation translates into the physical separation of individuals within a population into smaller sub-populations (demes). Eventually, organisms become reproductively isolated, and gene flow is prevented between demes. They are unable to interbreed and produce fertile offspring. When this happens, organisms that have been isolated by geographical isolation have developed into new species through mutation and natural selection.

- Allopatric speciation is the development of a new species when the population (demes) becomes physically isolated, occupying different geographical areas.

- Geographical isolation is the only factor to upset the Hardy-Weinberg law. It prevents random mating and reduces population size.

Intrinsic isolation

- Intrinsic isolation implies isolation by biological barriers that prevent interbreeding and the exchange of genetic material between individuals of different species.
- Sympatric speciation occurs when the population (demes) is reproductively isolated in the same geographical area or occupies geographically overlapping areas.
- Intrinsic isolation can be classified into : Prezygotic mechanisms (barriers) and Postzygotic mechanisms (barriers).

(a) Prezygotic barriers, which prevent mating or the formation of hybrid zygotes.

Examples

- Ecological isolation/habitat isolation
- Temporal/seasonal isolation
- Behavioral isolation
- Mechanical insulation
- Isolation of gametes, ...

(b) Postzygotic mechanisms, which reduce the viability or fertility of hybrid zygotes (preventing hybrids from developing or reproducing, even if fertilization is successful).

Examples

- Non-viability of the hybrid (fertilization occurs but the zygote fails to divide and develop).
- Hybrid sterility: the F_1 hybrids produced are infertile hybrids, unable to produce viable gametes.
- Degraded hybrids (F2 hybrids or backcrosses) have reduced viability or are infertile.

2.3.3.2 GENETIC DRIFT OR THE SEWALL WRIGHT EFFECT.

- Genetic drift is the variation in allele frequency over successive generations in small or isolated inbred populations due to random change rather than natural selection.

- In a small population, over time, this can lead to a drastic reduction in alleles or an increase in the frequency of certain other alleles.

- In a small population, this can lead to the extinction (increase in harmful alleles) of the population or, as the population becomes more divergent, the formation of new species.

- Genetic drift can occur in any population, but is more frequent in situations with **founder** and **bottleneck effects**.

2.3.3.2.1 THE FOUNDER EFFECT

The **founder effect** is the genetic drift that occurs when a small population is isolated or dispersed from the larger main population.

The smaller the population, the greater the effect of genetic drift. The founder population (new population) is started from a small number of pioneer individuals likely to have a gene pool that is not a representative selection of the original population. Thus, there may be a batch of certain alleles from the new population. Sometimes, rare alleles that are present at low frequencies in the original population may now be at higher frequencies in the gene pool of the new population.

The resulting barriers in the separation of populations can be physical or behavioral.

Continued reproduction in the pioneer population produces changes in allele and genotype frequencies for adaptation to different environmental conditions. The pioneer population becomes a new species and cannot reproduce with members of the original population.

Here are some examples of foundational effects in nature:

- The Baikal **seal** *(Phoca siberica) is* a species of freshwater seal that evolved in the isolation of Siberia's Lake Baikal. All of today's seals are descendants of a founding population of saltwater seals thought to have been lost when a sea passage linking Lake Baikal to the Arctic Ocean was cut off.

- A new population of self-pollinating plants resulting from the dispersal of a single seed.

- Dispersal of organisms from continental islands, such as Galapagos tortoises inhabiting the Galapagos Islands.

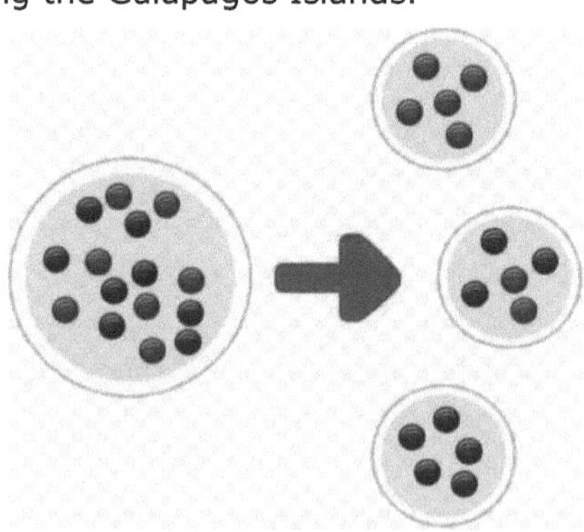

Figure 2.1 *Founder effect*

2.3.3.2.2 BOTTLENECK EFFECT

This is an example of genetic drift, which occurs when there is a drastic and sudden decline in a population due to unfavorable environmental factors.

- Natural disasters such as earthquakes, floods, droughts, fires, epidemic diseases and human activities such as hunting, logging and farming can considerably reduce the size of a population.

- A large number of individuals are killed non-selectively. The small surviving population has no gene pool representative of the large original population.

- This change, and the resulting loss of genetic variability, is known as the bottleneck effect.

- Most of the different allele types of the original population have been lost.

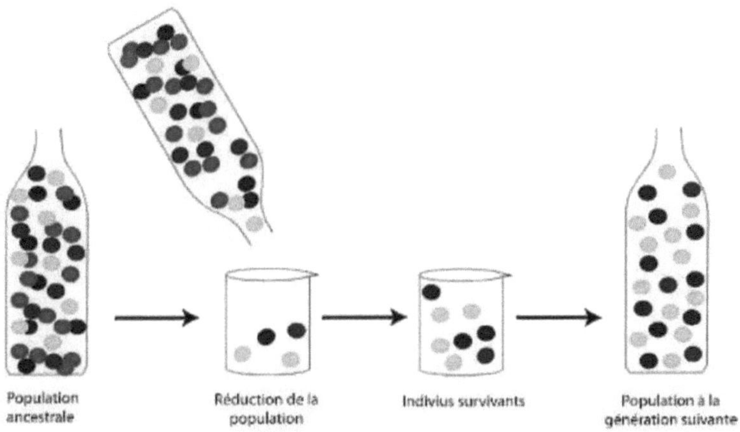

Figure 2.2 *Bottleneck effect*

After the population has gone through a period of considerably reduced size, there is a decrease in genetic variation in the population even though it later increases in size. This is because the mutation that causes population variation is a very slow process. It takes a long time and several generations for variation to increase in a population through mutation.

Examples:

- Man hunted northern elephant seals almost to extinction in the 19th century. The entire population was reduced to a maximum of 20 seals on Guadalupe Island. Due to the bottleneck effect, only a few genotypes contribute to the offspring of the next generation.

- Cheetah species have suffered a severe bottleneck effect due to human hunting, disease, lack of food or severe drought. Today, there are only three wild populations of cheetah, with very low genetic variability within these populations.

2.3.3.2.3 GENETIC FLOW

Gene flow is the migration or mating of individuals between populations that causes a corresponding movement of alleles. This can lead to changes in allele frequency.

Gene flow can increase variation within a population by introducing new alleles produced by mutation in other populations.

Continuous gene flow between populations tends to make gene pools more similar between populations. This reduces the diversity among populations on which natural selection can act. Gene flow between populations can prevent speciation.

2.3.3.3 HYBRIDATION

Hybridization is the process of obtaining one or more hybrid offspring by crossing or mating genetically different parents.

In **allopolyploids**, the extra set(s) of chromosomes come from more than one species. The F1 hybrid is generally sterile. If the chromosome complement is doubled, homologous chromosomes can be paired in meiosis. A fertile allopolyploid is produced.

2.3.3.4 ADAPTIVE RADIATION

Adaptive radiation or **divergent evolution** is the emergence of different forms from a common ancestor, dispersed to new inhabitants and diverse food sources, and their adaptation to live in a new environment.

Selective pressure causes changes in allele frequency. Over time, progressive modifications have taken place to transform certain body structures into different structures specialized for specific functions in new habitats. Eventually, these dispersed populations cannot reproduce with each other and with the original population. They have to become new species.

Examples:

Charles Darwin suggested that the ancestor of the **finches** had dispersed to the Galapagos Islands on driftwood or blown from the continent by strong winds. The finches' geographic isolation led to **allopatric speciation.**

Differences in beak size and shape show an adaptive radiation to different ecological niches. They are dull in color, and successful mating or recognition is achieved through courtship displays such as singing or other behavioral patterns.

Figure 2.3 *Darwin's Finches*

NOTES!

- The species is a category used in the classification of organisms. It is the lowest rank in the classification hierarchy.

- The same species can sometimes be subdivided into subspecies, races, varieties, demes and clines (progressive change of phenotype).

- Biologically, humans are classified as *Homo sapiens.*

- Interbreeding in the wild can be strongly influenced by unstable environmental variables and factors.

- Donkeys and horses belong to different species; donkeys can interbreed with horses to produce hybrids that are **not fertile** and cannot **interbreed.** A **mule** is the offspring of a male donkey and a female horse. The offspring of a male horse and a female donkey is called a **bardot**.

- A new population produced by the **bottleneck effect** has a **drastic reduction** in **genetic variability**. These individuals face a greater threat of new diseases, as they may not possess resistance against some of these diseases.

- **Convergent evolution** is the development of superficially similar structures or organs in unrelated organisms, because they live in a **similar environment**. Examples are the wings of birds and insects, and the streamlined bodies of fish and birds.

- Do not confuse **genetic drift** and **gene flow**: genetic drift is a change in the gene pool of a small population due to random change, or gene flow is the gain or loss of alleles from a population through the movement of individuals or gametes into or out of the population.

REVISION QUESTIONS

1. What is a variation?
2. Explain the 3 main sources of different variations within a population.
3. Name 5 main causes of variation in sexual reproduction.
4. What differences do you find between qualitative and quantitative variation?
5. What is the relationship between variation and adaptation in living species?
6. What is selection?
7. What's the fundamental difference between natural selection and artificial selection?
8. Describe the 5 main types of natural selection.
9. Explain the difference between an equilibrium polymorphism and a transient polymorphism.
10. What is hybridization?
11. What is a sperm bank?
12. Explain these concepts: homogamy, inbreeding, species, speciation, isolation, allopatric speciation, sympatric speciation, founder effect, bottleneck effect, gene flow, adaptive radiation, allo- and autopolyploidy, convergent evolution and divergent evolution, genetic drift.

CHAPTER 3: ARGUMENTS IN FAVOR OF EVOLUTION

For evolutionists, arguments in favor of evolution can be drawn from paleontology, comparative embryology, comparative anatomy, the geographical distribution of living beings, comparative biochemistry or DNA homology.

3.1 PALEONTOLOGICAL ARGUMENT

3.1.1 DEFINITIONS

Paleontology is the study of **fossils**, which are the remains of animals or plants more or less well preserved in sedimentary rocks or well buried.

Figure 3.1 *Remnant of a leaf in a rock*

These are the remains (or vestiges) of organisms that lived long ago.

Comparing fossils with present-day animal and plant organisms enables us to pinpoint the geographical and climatic conditions of the living environment, and to obtain **facies fossils**.

3.1.2 CHRONOLOGY OF THE LIVING WORLD

Fossils and ancient flora can be used to classify the terrestrial layers that contain them chronologically (**stratigraphy**).

The earth formed at the same time as the whole solar system, 5 or 6 billion years ago, the solid crust around 4 billion years ago, and the first living beings, which are prokaryotes - blue-green algae (cyanophyceae) and bacteria - appeared around 3 billion years ago. These earliest known fossils **are microfossils**.

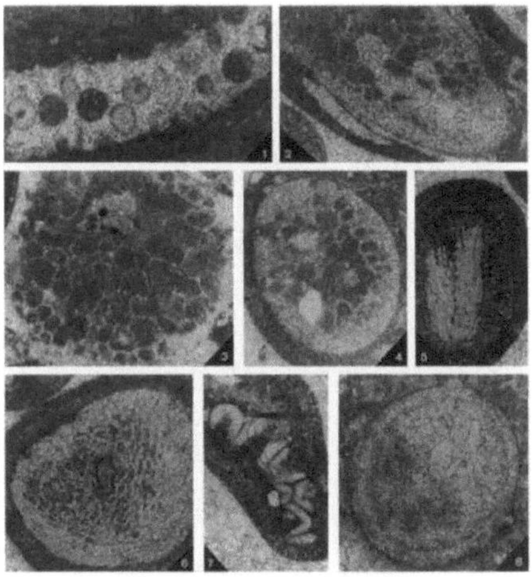

Figure 3.2 *Microfossils*

Microfossils are fossilized organisms of small size (mainly less than a millimeter), which can only be observed using special tools such as electron microscopes. They differ from **macrofossils**, which can be observed with the naked eye.

This remote period, corresponding to the first **fossilized** rocks, is known as the **Precambrian**. After the Precambrian, the earth's layers are divided into 4 **eras**, with each era itself subdivided into a number of **periods**.

Table 3.1 Chronology of the living world

ERES GEOLOGICAL	PERIODS	AGES/ YEARS	10^6 LINE ANIMALS	LIIGNEES VEGETALES
Quaternary	Holocene			
or			Men	
ANTHROPOZOIC	Pleistocene	3,3		
Tertiary or	Pliocene		Mammals	
CENOZOIC	Miocene		current	
	Oligocene			
	Eocene			
	Paleocene	65		
Secondary	Cretaceous			Angiosperms
or		136		
MESOZOIC	Jurassic		Birds	
		150	primitives	
	Triassic		Mammals	
		225	archaic	
Primary	Permian			Gymnospremes
or		280		
PALEOZOIC	Carboniferous		First	
			reptiles	
			insects	
		345		
	Devonian		Amphibians	Fougères
				Mushrooms
		395		
	Silurian		Poisons	foams
		430	archaic	
	Ordovician		Agnathes	
		500		
	Cambrian		Prevertebrates	

	570	
PRECAMBRIEN	Crustaceans Molluscs Echinoderms Annelids Coelenterates Sponge Protozoa	Algae Eukaryotes

3.1.3 STROMATOLITES

We can't discuss life in the Precambrian without mentioning ***Stromatolites***, which are structures built by bacteria and or blues algae (prokaryotes). These are rocks formed by bacteria that encrust with sediments suspended in water.

Stromatolites are fossilized carbonate structures. They are among the oldest known fossils, and certainly the oldest macrofossils; they have been known since 3.5 Ga (in Australia).

Stromatolites therefore played an essential role in **the evolution of life** on Earth, enabling the development of aerobic life (needing oxygen). This in turn enabled the development of more complex, multi-cellular, terrestrial organisms such as humans.

Figure 3.3 *Stromatolites.*

3.1.3 THE 5 MASS EXTECTIONS

Mass extinction is a subject of relative truth, since a new discovery in geology or paleontology could turn everything upside down and change our certainties.

Generally speaking, during a mass extinction, an average of **75%** of the animal and plant species present on earth and in the oceans disappear. There are five mass extinctions: the Ordovician, Devonian, Permian, Triassic-Jurassic and Cretaceous-Tertiary.

Here are biology'**s five major crises**:

Ordovician-Silurian **extinction**. 450 million years ago.

Devonian **extinction**. 375 million years ago.

Permian Triassic **extinction**. 250 million years ago.

Triassic-Jurassic **extinction**.

Cretaceous-Paleogene **extinction**.

Table 3.2 *Geological eras and major biological extinctions*

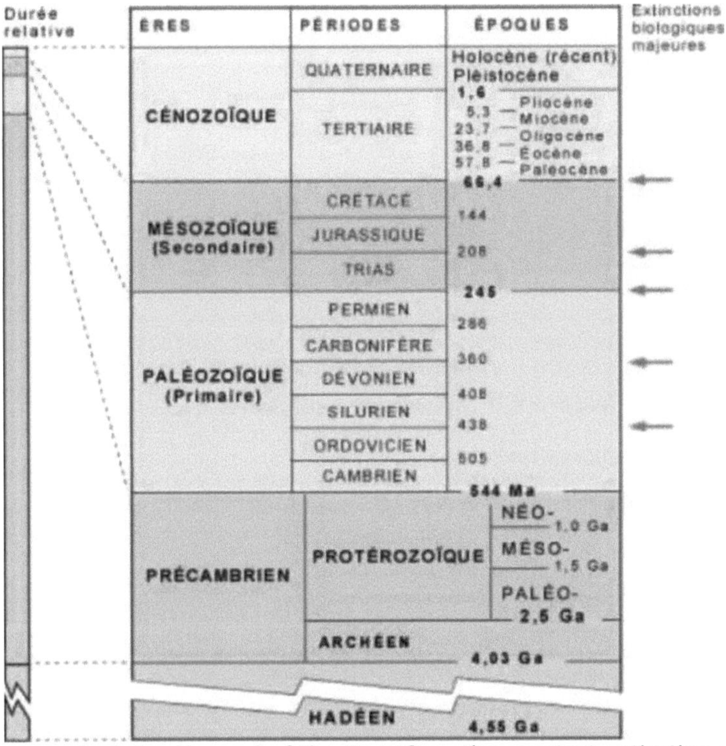

This group includes the end of the **Permian**, the greatest extinction of all time around 252 million years ago, responsible for the disappearance of 95% of marine species.

The **five** (hypothetical) **causes** of these extinctions have now been identified: changes in land and sea use, direct exploitation of certain organisms, climate change (intense ice age, volcanism), pollution, invasive alien species and meteorites.

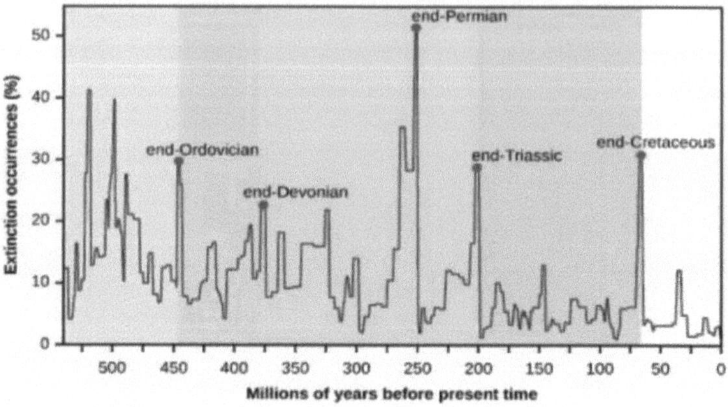

Figure 3.4 *The 5 major biological extinctions*

3.1.5 GREAT OXYGENATION

The **Great** Oxygenation Event (GOE), also known as the **oxygen catastrophe** or oxygen crisis, is a major event that took place in the Earth's atmosphere, surface waters and biosphere around 2.4 to 2 billion years ago (Ga).

In fact, almost all the Earth's breathable oxygen (nearly 21% of the Earth's atmosphere) came from the oceans. It accumulated in the atmosphere thanks to marine micro-organisms (e.g. cyanobacteria and planktonic micro-algae) capable of photosynthesis.
For many **anaerobic** organisms, oxygen is not only **useless**, it's **toxic**. It's a poison that rapidly destroys them. For them, oxygen is synonymous with **death**.

2.4 billion years ago, a mass extinction, dubbed the Great Oxidation, killed off most existing organisms, for whom the oxygen then released into the atmosphere was **toxic**.

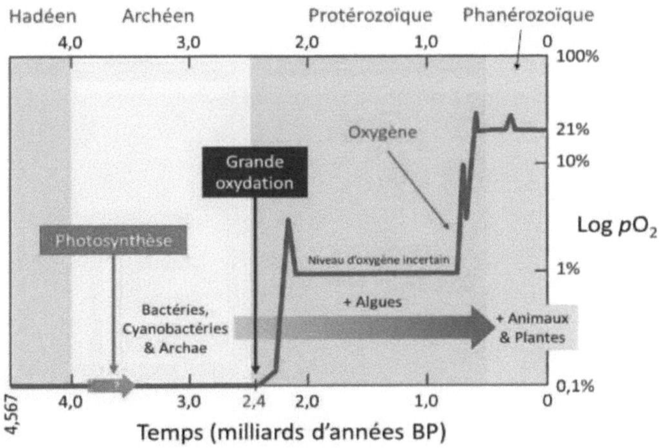

Figure 3.5 *The major oxidation stages*

3.1.6 EXAMPLES OF MASS EXTINCTION

A. Trilobites

Trilobites *(Trilobita) are* a class of fossil marine arthropods that existed during the Paleozoic (Primary Era) from the Cambrian to the Permian.

Their disappearance is probably due to their inability to compete with, or escape from, active and combative **fish**. Other trilobite predators include starfish, cephalopods, coelenterates, etc. Fossils.

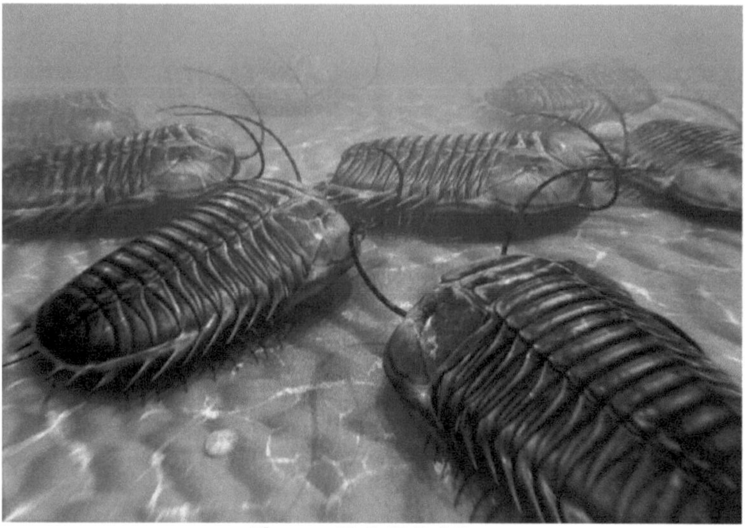

Figure 3.6 *Trilobites*

B. Ammonites

Ammonites are a group of fossil marine animals in the cephalopod mollusc class. The shell is generally the only fossilized part. It takes the form of a flat spiral with contiguous turns.

These are cephalopod molluscs. In nature today, these tentacled organisms include octopuses, squid, cuttlefish and nautiluses.
However, victims of the crisis that decimated the fauna at the end of the Cretaceous 65 million years ago, ammonites have now completely disappeared. Ammonites, which are exclusively marine animals, inhabited continental seas and, occasionally, oceans for almost 325 million years.

Figure 3.7 *An ammonite fossil*

3.1.4 INTERMEDIATE FORMS

L eontology has discovered that some fossils have characteristics of both classes. These fossils are called intermediate forms.

- **Ichtyostega** is intermediate between fish and amphibians.
- **Seymouria** ranks among amphibians and reptiles.
- **Therapsids**, intermediates between reptiles and mammals.
- **Archaeopteryx**, between reptiles and birds.

Figure 3.8 *Archaeopteryx fossil*

3.1.5 HORSE DEVELOPMENT

To support the evolutionary thesis, scholars often cite the Equidae series, and sometimes the Camelidae and Proboscidae series.

Here are the characteristics of some of the many forms of Equidae that lived during the Tertiary Era

EOCENE

Second Tertiary period, around 60 million years ago.

Type: *Eohippus* or *Hyracotherium.*

Size: 30 to 40 at the shoulder, like a dog. Legs are short, ending in 3 or 4 separate toes, each bearing a small hoof.

The small, numerous teeth lack the sharp ridges of herbivores. They are covered with small tubers, like omnivores.

Eohippus is an animal adapted to life in the forest, feeding on the leaves of shrubs and moving on hard ground thanks to its separate toes.

OLIGOCENE

Several types of Equidae are descended *from Eohippus,* for example *Mesohippus,* with the following modifications:

The animal has grown in size, measuring around 60cm at the shoulder, like a sheep.

Leg elongation occurs mainly in the middle finger (3ème finger), which becomes longer and, above all, stronger than the other fingers. The latter remain slender and are no longer functional.

The skull lengthens. Teeth widen and bear a few sharp ridges characteristic of the herbivorous diet.

MIOCENE

With a drier, deeper climate, forests are replaced by hard grasses. Horses became runners and herbivores. Take, for example, the case of *Merychippus*, almost the size of today's horse.

The legs have become very long, with the whole body resting on the 3rd toe, which ends in a sturdy hoof. The other toes are only present in traces.

The head is long with high, straight teeth whose continuous growth compensates for wear. Teeth are covered with enamel, typical of herbivores.

PLIOCENE

At that time, there were many different types of Equidae. They hardly differ from today's *Equus cabalus.* The lateral toes have almost completely disappeared.

Many of these genera became extinct during the Pleistocene to Quaternary Ice Age, with the exception of those that migrated to

Eurasia and North Africa, giving rise to zebras, donkeys and wild horses.

In **conclusion,** this evolution would be accompanied by a reduction in the number of fingers, an increase in height, a change in dentition and diet.

Paleontologists have reconstructed the evolution of other zoological groups, such as **camelids** (camels) and **proboscideans** (elephants).

Table 3.3 *Equid evolutionary history*

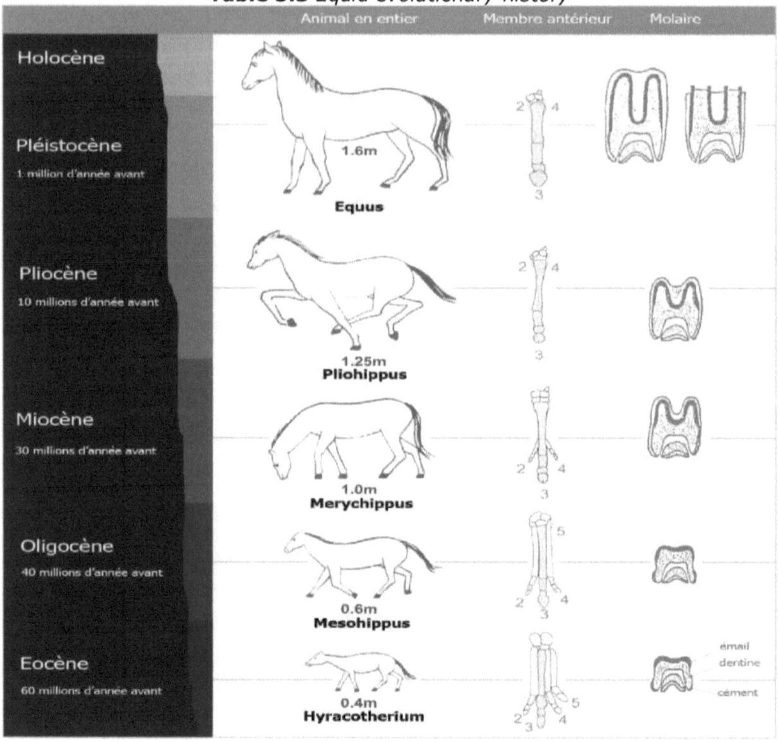

3.1.6 HUMAN PALEONTOLOGY

3.1.6.1 SPECIAL CHARACTERISTICS OF MAN

In anatomy and physiology, man is close to the anthropoid primates: Orangutan, Chimpanzee and Gorilla. But he differs from them in many respects, which it would be impossible to list here:

- A perfect upright position.
- A face-lift.
- Three times the cranial capacity of today's anthropoids.
- Extreme hand mobility with a thumb that can be opposed to the other fingers...

Man **differs** from anthropoid primates in many other psychological and sociological ways. Above all, the ability to create tools. These tools always accompany human remains, enabling us to assess the state of hominization.

3.1.5.2 PRE-HUMANS

Man is not a descendant of the ape; anthropoids and humans are separate and distinct branches. The general opinion of specialists is that the origin of the lineage that led to man is very old, dating from the second half of the Tertiary era, with an age of around 20 to 10 million years.

We don't know what tertiary man was, but the great tertiary primates resemble man much more than they do today's anthropoids. This large primate, called **Areopithecus**, had humanoid features (teeth, pelvis, biped...) but probably only represents a small branch parallel to the human branch.

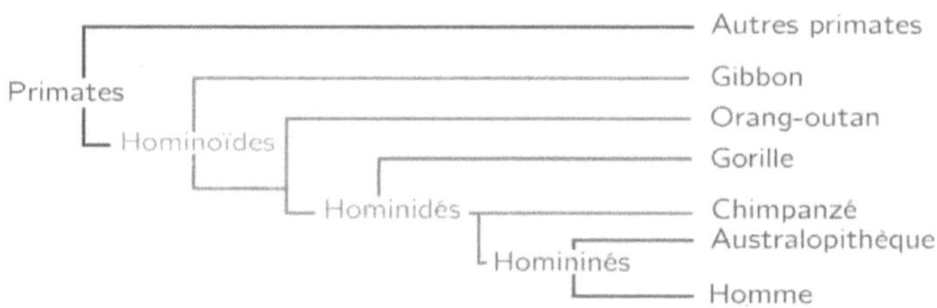

Figure 3.8 *The evolution of Primate paleontology*

3.1.5.3 THE STAGES OF HOMINIZATION

3.1.5.3.1 Australopithecus and *Homo habilis*

The oldest fossils belonging to the humanoid lineage were found on the eastern edge of Africa, first in the south, hence the name **Australopithecus**, then more recently in Kenya, Tanzania and Ethiopia.

Australopithecans can be divided into two basic forms: the so-called **gracile** form (without sagittal ridges) and the **Paranthropes** (South African Paranthropes, Tanzanian Zinjanthropes, Ethiopian mandibles), all with sagittal ridges.

Australopithecans still possess a number of **simian characteristics**, including a small brain volume of between 400 and 600 cm^3 , and a prognathic face, although all these features were less pronounced than in Anthropoids.

In other words, Australopithecines were **human**: the occipital orifice was on the vertical plane, the dentition was humanoid, standing ...

What's more, Australopithecines were already craftsmen of a **crude Iithic industry**. They also used bone to make tools.

Four species of Australopithecus are known:

- *Australopithecus afarensis* or Pre-australopithecus
  Known since: 5 MA, most recent: 2.7 MA
  Size: 1.10 to 1.30 m
  Cranial volume:300 to 400 cm3
  Biped
  Location: Africa
- Australopithecus africanus
  Known since 3 MA, most recent: 2.2 MA
  Height: 1.30 m
  Cranial volume: 400 to 600 cm^3
  Biped
  Location: Africa
- Two other species: *Australopithecus boisei* and *Australopithecus robustus.*

In the same deposits as the Australopithecines, fossils of an ancient human type have been discovered, associated with **cut pebble** tools. These fossils probably represent the earliest form of the **Homo** genus, dating from 2 to 1.4 million years ago. The presence of these early tools earned this type of hominid the name Homo *habilis*.

Characteristics of **skilled man** or *Homo habilis* :

- Known since: 2MA, most recent: 1.5 MA
- Height: 1.20 m
- Cranial volume: 600 to 770 cm^3
- First tools: shaped pebbles
- Location: East Africa

All researchers agree that *Australopithecines* **belong to the human lineage**. Along with *Homo habilis*, they represent **the first stage of Thominization**.

This very distant phase remains poorly understood and the subject of much debate. On the other hand, we know more about the hominids that lived from 1 to
1,3 million years.

For the sake of simplicity, we distinguish 3 successive phases in the formation of present-day man: the Pithecanthropians, the Neanderthals and the *Homo sapiens.*

3.1.5.3.2 Pithecanthrope or *Homo erectus*

These include the Pithécanthropes of Java, the Sinanthropes of Peking and Shensis (China), the Atlanthrope of Ternifère near Oran, the Rabat and Casablanca Men, the Telanthropes of South Africa, certain Oldoway Men (Tanzania) and the Chadanthropes.

They are characterized by these elements:

- Known since 1.6 MA, the most recent: 150,000 years ago
- **Cranial volume** increased by 800 to 1200 cm^3
- **Bifacial stone** tools and,
- Conquering **fire**.
- Location: Africa, Europe and Asia (China).

Pithecanthropes can be distinguished from present-day humans mainly by their more powerful mandibles and smaller cranial volume.

3.1.5.3.3 Neanderthals or *Homo sapiens Neanderthalensis*

Neanderthals represent a much more recent evolutionary stage than Pithecanthropians. They appeared in **Eurasia** during the last WURN glaciation, around 70,000 years ago.

He was no taller than 1.55 m, with a very large skull, larger than that of *Homo sapiens*: around 1,600 cm^3 . Flint tools became more sophisticated and diversified. Some skeletons were intentionally buried, the first known evidence of **burial**.

3.1.5.3.4 Present-day man or *Homo sapiens sapiens*

Known as modern man, it existed 100,000 years ago. Location, worldwide. His skull (1300 to 1500 cm^3) is different from that of the Cro-Magnon type (1600 to 1700 cm^3).

Tools had made great progress, and **sedentary life** only began with **agriculture** in the **Neolithic** period.

In **conclusion**, all these human forms fit together so perfectly that the **authors cannot agree** on the nature of the Primates of the human lineage.

These stages of human evolution do not correspond to a single, simple lineage; rather, Australopithecines, Pithecanthropes and Neanderthals form a thicket of branches tending towards three structural stages.

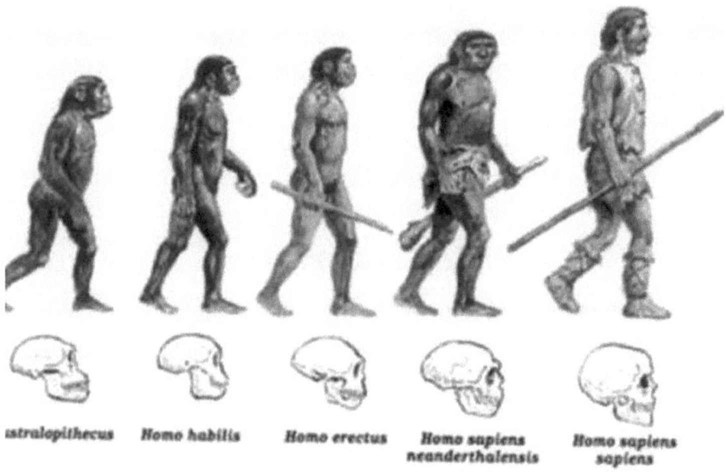

Figure 3.9 *Ihominization stages*

Hominization took place in several parts of the world, producing several different branches, only one of which has persisted, evolving into present-day man, because, for biologists, all the current human species comes from a single lineage and from a **single** common **ancestor**: this is **monogenism**, which is in opposition to **polygenism**.

3.2 COMPARATIVE ANATOMY

Widely dispersed groups that originate from a common ancestor will share certain basic structural characteristics

The degree of similarity between these groups indicates the degree of closeness to the common castor bean.

Comparative anatomy is used as evidence to establish the evolution of any relationship on the basis of structural similarities or differences.

3.2.1 HOMOLOGOUS STRUCTURES

(a) **Homologous structures** are modified structures that share a basic model as a common ancestor, but may have different forms and functions. This enables the descendant to occupy ecological niches.

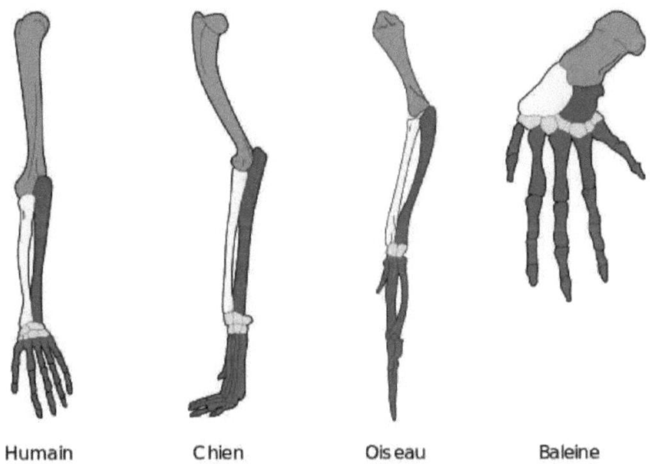

Figure 3.10 *Homologous structures*

(b) This is known as **divergent evolution**.

(c) **An example** of a homologous structure is the pentadactyl limb. The basic pentadactyl pattern is found in amphibians, reptiles, birds and mammals. Divergent evolution has led to the modification of limbs for different uses, resulting in **adaptive radiation**.

3.2.2 SIMILAR STRUCTURES

(a) Analogous structures are structures that perform similar functions but do not derive from a common ancestor.

(b) This is called convergent evolution. Organisms (or their body structures), because of a similar environment, are more similar than their ancestors (or ancestral structures).

Examples:

Grasshopper and mammal legs.

Butterfly and bird wings.

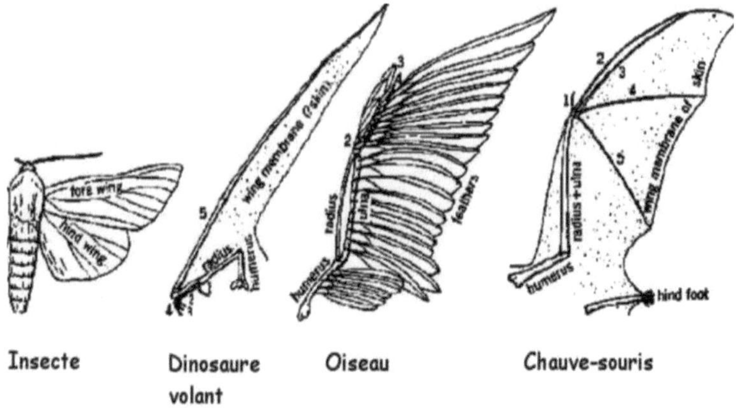

Figure 3.11 *Analog structures*

3.2.3 VESTIGIAL ORGANS

Vestigial organs are remnants of structures that had important functions in their ancestors. They have diminished in size because their functions have diminished or they are no longer needed.

Some examples of vestigial organs:

(a) The **cecum**, **appendix** and **coccyx** in humans.

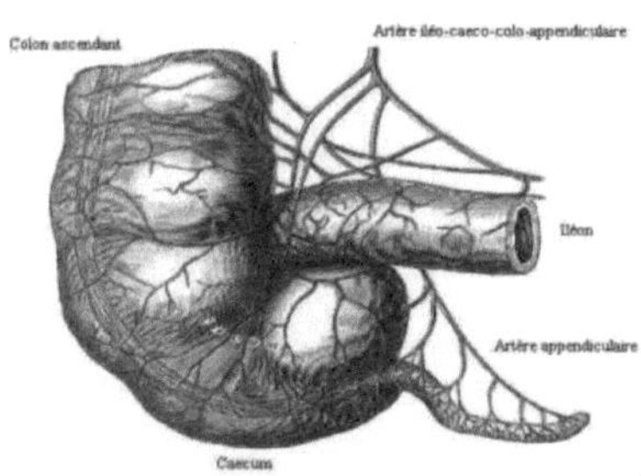

Fig. 3.12 *The human appendix*

(b) In some snakes, there are two small vestigial bones on each side, representing the **ilium** and **femur** of walking ancestors.

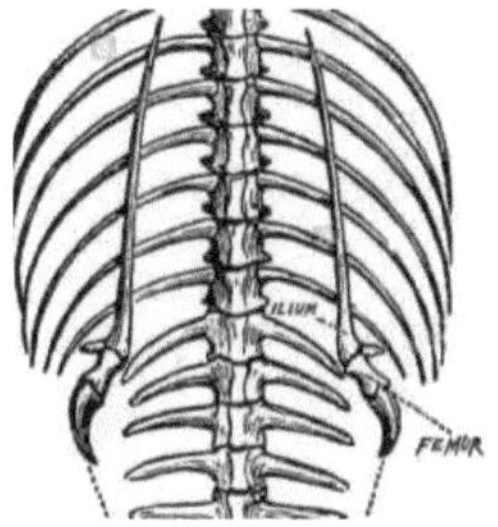

Figure 3.13 *Remnant of an ophidian femur*

(c) Remains of **pelvic bones** in whales that have lost their function.

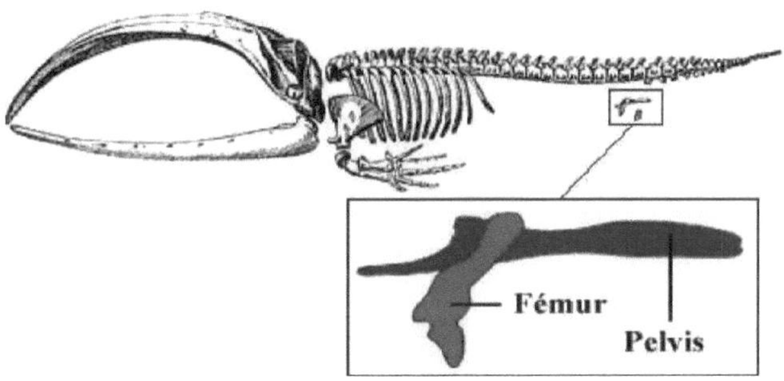

Figure 3.14 *Remnant of a whale pelvic bone*

3.3 COMPARATIVE EMBRYOLOGY

Comparative embryology, the theory of German naturalist Ernest Haeckel.

Ernest Haeckel
(1834-1919)

The early embryological development of vertebrate animals is similar and reflects evolutionary origins.

The embryo of a superior class, in whole or in part, passes through stages that reproduce embryonic states characteristic of systematically inferior animals.

Adult vertebrates, for example, may be quite different from one another, but they do have certain structural similarities, particularly during the embryonic stages of cleavage, gastrulation and organogenesis.

For example, all vertebrate embryos go through fish-like stages with pharyngeal clefts. In humans, however, these develop in the eustachian tubes and gills of fish.

This gives rise to the theory of **ontogeny** (development of the individual), **which recapitulates phylogeny** (development of the species or group). The early embryonic development of an individual reflects its evolutionary origins.

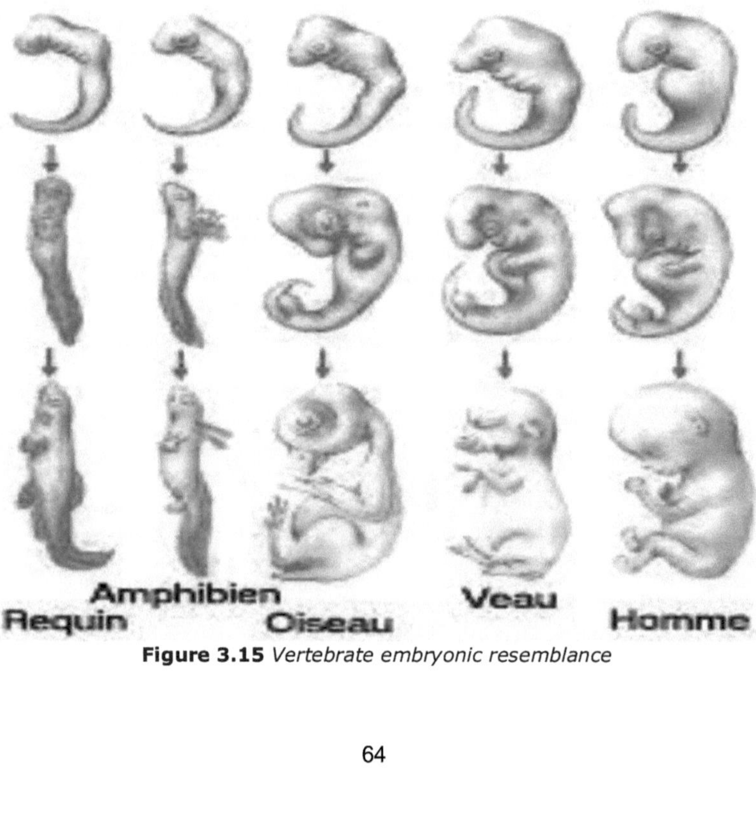

Figure 3.15 *Vertebrate embryonic resemblance*

Similarly, annelids, molluscs and plathelminths develop through a trochophore larva.

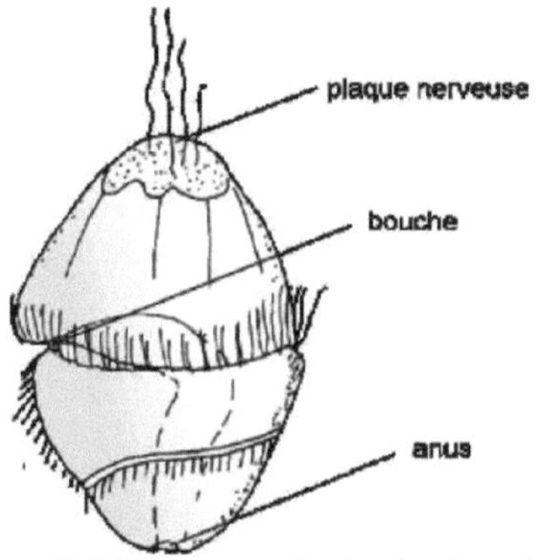

Figure 3.16 *Diagram of a trochophore larva*

3.3 GEOGRAPHICAL DISTRIBUTION

Biogeography is the study of the geographical distribution of species.

Plant and animal species are generally native to a particular shared space.

Millions of years ago, the continents slowly separated and changed shape.

Closely related plants and animals were dispersed from the area of origin when the continents separated. Organisms were then isolated by physical barriers such as the ocean, sea or mountains.

Selection pressures in different environments can lead the dispersed population to adapt to their respective new ecological niches. They gradually evolve into distinct species through **adaptive radiation**.

There are many examples of closely related plant and animal species found in widely separated parts of the world.

Examples

(a) Members of the camelid family, the camels of North Africa and Asia, and the "Llamas" of South America, are thought to have descended from a common ancestor in North America.

(b) Darwin's Finches in the Galapagos Islands.

(c) The different types of fish with **lungs** (Dipneustes) in Africa, South America and Australia (Lepidosiren, Protopterus and Neoceratodus).

(d) Monotremes and Marsupials

3.4 COMPARATIVE BIOCHEMISTRY

The study of chemical molecules in numerous organisms has revealed biochemical homologies.

The degree of similarity of biochemical homologies indicates the degree of evolutionary relationship.

One example is **cytochrome c**, where the exact sequence of amino acids in the polypeptide has been determined for different animals to show an evolutionary relationship.

Table 3.4 *Number of amino acid differences with human cytochrome c*

Organization	Number of amino acid differences with humans
Rhesus monkey	1
Rabbit	9
Cow	10
Pigeon	12
Drosophila	24
Yeast	42

Another example is the molecular data on human beta hemoglobin and its evolutionary relationship with other vertebrates. Animals with fewer differences in amino acid numbers are probably more closely related in evolution.

Table 3.5 *Number of acids of difference with the beta chain of human hemoglobin*

Organization	Number of amino acid differences from human hemoglobin polypeptide. Total chain length = 146 amino acids.
Rhesus monkey	8
Horse, cow	25
Mouse	27
Hen	45
Frog	67
Lamprey (fish)	125

It has been found that all living organisms use the same biological molecules for a similar function.

Examples

(a) DNA and RNA nucleic acids store and express genetic information.

(b) Most **enzymes** are made up of **globular proteins** and catalyze many similar types of biochemical reaction.

(c) ATP, carbohydrates and fats used for stored energy.

These suggest a common ancestry for all organisms.

A **molecular clock** is a method of evolutionary synchronization. It is based on the concept that, in the course of evolution, the number of substitutions in the nucleotides of nucleic acids (DNA or RNA), and therefore in the types of proteins encoded by nucleic acids, is proportional to time.

3.5 DNA HOMOLOGY

DNA is the genetic material present in the cells of all organisms. This shows that all organisms have similar origins.

The genetic code is universal: the same DNA code encodes the same amino acids in almost all organisms.

Another proof is to compare the DNA nucleotide sequence of genes.

The number of similar and different DNA gene sequences is an indication of the evolutionary relationship between organisms.

Another proof is to compare the DNA nucleotide sequence of genes. The number of similar and different DNA gene sequences is an indication of the evolutionary relationship between organisms.

The techniques commonly used to compare DNA homology are hybridization and chromosome painting.

(a) The DNA hybridization technique can be used to compare the base sequence of DNA nucleotides across the genome of different organisms.

(b) In chromosome staining, specific fluorescent probes are attached to targeted parts of chromosomal DNA. It can be used to identify interspecific homology. Fluorescent probes attached to chromosomal DNA from one species are used for hybridization with DNA from other species.

REVISION QUESTIONS

1. Explain these concepts: paleontology, stratigraphy, fossils, facies fossils, microfossils, macrofossils, geological eras, geological periods, intermediate forms.
2. What are stromatolites?
3. How have stromatolites changed life on Earth?
4. How were stromatolites formed?
5. What is a mass extinction?
6. What are the 5 great extinctions?
7. What are the 5 causes of species extinction?
8. What is the greatest mass extinction?
9. What is an oxygen catastrophe?
10. What's the main source of oxygen?
11. Is oxygen poisonous?
12. What were the consequences of the great oxidation?
13. What caused the disappearance of the trilobites?
14. What is a fossil ammonite?
15. Where is the ammonite now?
16. Name some of the arguments evolutionists use to support the evolution of living beings.
17. Explain the 3 major transformations observed during the evolution of equids in the Tertiary period.
18. Explain at least 5 characteristics that distinguish humans from anthropoid apes.
19. What is hominization? Describe the main stages of hominization.
20. Use an example to explain the difference between analogy and homology of anatomical structures.
21. How can you explain Haeckel's aphorism that "ontogeny recapitulates phylogeny"?
22. What is a molecular clock?

CHAPTER 4 EXPLANATORY THEORIES OF EVOLUTION

4.1 INTRODUCTION

The facts of adaptations, paleontological documents, embryological observations, comparative anatomy and other facts that fall outside the scope of our program, lead evolutionists to seek the best possible, most coherent explanation in evolution, the one that coordinates the greatest number of observed facts.

If evolution is a fact, it is above all a historical fact. It's one thing to observe that living beings have changed over countless generations; it's quite another to determine the causes of these variations.

If we try to go even deeper, we can ask ourselves how this evolution came about? It's these attempts at explanation that we refer to as the theory of evolution.

We'll look in turn at the main theories that have been built up, from reflections and research, which persist more or less in the form of trends in modern attempts to explain evolution.

4.2 LAMARCK'S THEORY

4.2.1 INTRODUCTION TO LAMRCKISM

Jean Baptiste de Monet Chevalier de **Lamarck** (1744-1829), Professor of Zoology at the Museum of Natural History in Paris, was the first to use the idea of **heredity of acquired traits**, more or less well presented by his predecessors (Aristotle, Hippocrates, De Maupertuis, Linné, Buffon...), to explain the transformation of living species.

Lamarck expounded his theory in the *opening speech of his zoology course* in the year VII, and refined it in his work entitled *Zoological Philosophy,* 1809. He is often referred to as the **father of** evolutionism.

Lamarck was commissioned to classify the collections of the Natural History Museum in Paris. He classified **invertebrates** into **10 groups**, and realized that the divisions between the different groups were often artificial. The transition from one species to another often seems gradual. **The study of fossils** also shows variation over time. These observations led him to assume a real, historical kinship.

Lamarck's theory thus encompasses the whole of nature, deriving minerals, animals and plants from a **common source**, which he refers to as **gelatinous** and **mucilaginous molecules**, under the influence of **electricity**.

4.2.2 LAMARCKIAN EXPLANATIONS

Lamarck draws mainly on facts of nature. He mentions, for example, the case of the **mole** and cave-dwelling animals. If these animals are blind, he says, it's because in the dark they don't need to use their eyes, which have gradually atrophied and no longer function.

The **Giraffe** has a long neck because it lives in places where the almost arid, grassless earth forces it to graze on the leaves of trees and continually strive to wait for them.

Such explanations may fit the **paleontological facts**: the horse's ancestors quickly had to cover large distances to flee enemies and search for food. **Over time**, the foot lifts, and the middle toe bears most of the body's weight; as a result, this toe grows stronger, and the unused side toes gradually disappear.

Lamarckian explanations also apply to **embryological facts** and other **adaptations**.

Figure 4.1 Lamarckian adaptation of the Giraffe to its environment

4.2.3 DOCTRINE: LAMARCKISM

The Iamarckian doctrine is based on two principles:

- **The adaptation of the organ to the living environment,** explained by the law of use and non-use: the environment imposes needs, and needs give rise to efforts and habits; habitual use of an organ strengthens it, and lack of use atrophies it and causes it to disappear. We translate this law by his aphorism: "the **need creates the organ"**.
- **Inheritance of traits acquired under the influence of the living environment**: for traits to become more pronounced, they must

be passed on to descendants from generation to generation.

4.2.4 DISCUSSION, CRITICISM AND OBJECTIONS

They raise their voices about one or other law (or principle):

The **environment** certainly acts on the organism, but function **does not create** an organ; it can adapt, transform or develop it. The law of use can only apply to an organ that already exists, it cannot create it. This law does **not explain the development of** troublesome organs, such as long horns that make walking difficult for some cows.

No experiment **has been able to** show **the heredity of acquired traits**. When an environment acts on non-reproductive organs, variations are not hereditary; they are **somations**. But when the environment acts on the reproductive organs, they are as random as **mutations**.

In the current state of knowledge, no case of heredity of acquired characters can be cited. Lamarck's great merit lies in bringing together in a single theory the 3 major elements of evolution:

- The succession of living beings (relays and extinction).
- A long duration allowing it.
- And a perfecting of the living world, from the simple to the complex, from the less perfect to the more perfect.

Lamarck's theory is now only of **historical** interest.

4.3 DARWIN-WALLANCE THEORY

4.3.1 THE VOYAGE ABOARD THE HMS BEAGLE

In 1832, Charles Darwin (1809-1882) was the British naturalist on a five-year tour of many countries, including the east coast of South America, the Galapagos Islands, New Zealand, Tasmania and South Africa. Reading Lyell's "Progress in Geology", he extended Lyell's method and conception to biology. He brought with him a wealth of documentation.

Darwin had also been struck by English economist Thomas Malthus's 'Essays on the Principle of Population'. Malthus's ideas, according to which individuals multiply in geometric progression while the quantities of food available increase only in arithmetical progression, had a profound influence on Darwin and are at the origin of the notion of the struggle for life.

Note: **HMS** is a Royal Navy ship, Her Majesty's Ship or His Majesty's

Ship.

He studied a wide range of flora and fauna he'd never known existed. He examined fossils found in rocks and discovered that different forms of life had undergone numerous changes. On the volcanic islands of the Galapagos, Darwin discovered new species of flora and fauna, different from those on the mainland.

When he returned to London, for many years Darwin was reluctant to publish his ideas. He was finally persuaded by Alfred Wallace, another naturalist who had also studied flora and fauna during his travels in other countries. Darwin and Wallace jointly presented their paper "The *Theory of Evolution and* Natural *Selection"* to the **Linnaean Society** in 1858.

The following year (1859) Darwin published his book on *The Origin of Species by Means of Natural Selection,* supported by the extensive evidence he had observed during his voyage on the HMS Beagle.

4.3.2 THE DOCTRINE: DARWINISM

Charles Darwin and Alfred Wallace proposed that new species arise from pre-existing species through the mechanism of natural selection.

The theory is based on **three observations** and **two deductions**.

- **Observationl**: Overproduction of offspring

Organisms produce more offspring than are needed to replace the parents.

- **ObservationZ**: relative constancy of numbers

The number of individuals in the natural population tends to remain stable in size over long periods.

- **Deductionl**: Struggle for existence

There is competition for survival (struggle for existence) within a population. Many individuals fail to reach maturity and reproduce.

- **Observations**: a variation exists within the population

Variations exist between different individuals of a given species. Darwin distinguishes between **definite variability**, which occurs in the same way in all modified individuals, and **indefinite variability**, which is random and changes from one individual to the next (mutations).

- **Deduction?** Survival of the fittest through natural selection.

In the struggle for existence, individuals with the variations best adapted to their natural environment operating at the time will be selected for survival. They have a reproductive advantage and can pass on their characteristics to their offspring. Over time, the same species can give rise to distinct groups that are sufficiently distinct to belong to different species.

4.3.3 EXAMPLE OF A GIRAFFE

Taking the giraffe as an example, according to the Darwin-Wallace theory, there was a variation in the neck length of the giraffe's ancestors. During periods of food shortage, leaves from the lower branches of trees had been eaten. Only giraffes with longer necks could reach the highest leaves. Natural selections would favor long-necked giraffes, and these would survive by passing on their characteristics to their offspring. Short-necked giraffes would starve.

Figure 4.2 *The example of the giraffe: Lamarck's need creates the organ, and Darwin's survival of the fittest.*

4.3.4 CRITICISMS AND OBJECTIONS

Darwin's theory initially met with great success, but it was soon realized that it raised a number of objections and encountered considerable difficulties.

Natural selection, as conceived by Darwin, does not correspond to what happens in nature. Natural selection is more a factor of stability than evolution, favoring the average type.

The small individual variations Darwin refers to are somations, not hereditary. Darwin and Wallace were unaware of the mechanism of genetic inheritance.

Darwinism admits variability within a species, but doesn't explain it. It's hard to understand how, from these small variations, we can arrive at entirely new organs like bird feathers?

Darwinism, in its old, classical form, has collapsed. Like Lamarckism, its place in the history of biology remains significant.

4.4 THE DE VRIES THEORY

4.4.1 OBSERVATIONS

Hugo de Vries was observing a plant, the Œnothera or Evening Primrose *(Œnothera lamarckiana),* cultivated in the laboratory near Amsterdam, when he suddenly saw new varieties appear, with the characteristics of these varieties being transmitted in their entirety to their descendants. He called these sudden hereditary variations mutations.

4.4.2 DOCTRINE: MUTATIONISM (1900)

For De Vries, in his book ***Die Mutationstheorie***, mutations are the only variations recognized as hereditary; they alone have evolutionary value.

In some cases, abrupt variations have led to the creation of new species. Evolution was no longer progressive.

4.4.3 CRITICISMS AND OBJECTIONS

Mutationism is a long way from explaining all the evolutionary characteristics of the living world.

Mutations are **fortuitous**, **accidental** and **indifferent**, and therefore have no evolutionary significance in themselves.

They are generally **subtractive**, never creating a new organ, and result not in the acquisition but in the deletion of good characters.

Mutations **are low-amplitude variations** that never go beyond the scope of the species.

4.5 THE SYNTHETIC THEORY OF EVOLUTION

4.5.1 THE NEED FOR A SYNTHETIC THEORY

It was in the early 1940s that a **comprehensive theory** of evolution was developed, known as the synthetic theory of evolution. It is said to be synthetic because it integrates discoveries and principles from many fields.

The authors of modern synthetic theory include :

- British statistician and biologist R.A. Fischer (1890-1962), who demonstrated Mendel's rules for the transmission of hereditary traits.

- British-born Indian biologist J.B.S. Haldane (1892-1964), who studied the rules of natural selection.
- Other scientists who contributed to the development of synthetic theory include Russian geneticist Theodosius Dobzhansky (1900-1975) and American geneticist Sewall Wright (1889-1988), German biogeographer Ernst Mary (1904-2005), American paleontologist George Gaylord Simpson (19021984), and American botanist G. Leydard Stebbins (1906-2000).

Synthetic theory involves so many factors that it's difficult to interpret its exact scope, but its broad outlines are clear.

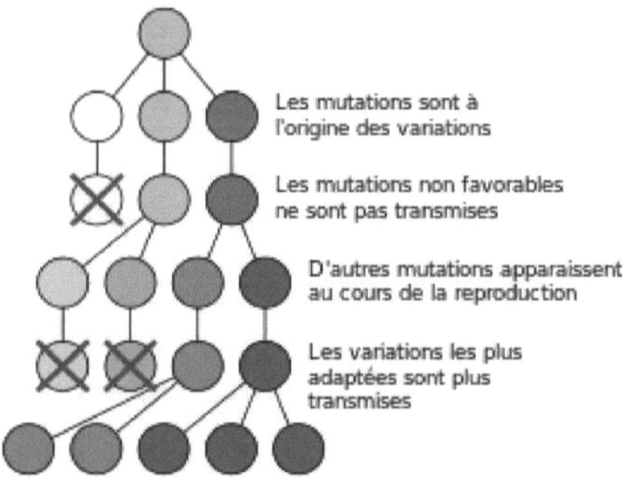

Figure 4.3 *Mutations and natural selection*

4.5.2 MUTATIONS

Mutations account for the formation of new types, and are a real **source of variability** within a population.

Although most of them have a negative effect, some may have one or other advantage in a given environment. These mutations occur entirely at **random**, and are subject to the control of **natural selection**.

4.5.3 NATURAL SELECTION

Natural selection, as a **factor in the elimination** of the **least suitable** genotypes, accounts for adaptation to the living environment.

4.5.4 ENVIRONMENTAL ACTION

Natural selection is nothing other than the action of the environment on living beings. The environment directs selection in the direction of better adaptation. The environment tends to progressively eliminate unfavorable mutations, retaining only those that provide any advantage to the individuals carrying them.

Selection is **conservative** if the environment is constant; it is **innovative** in response to any change in the environment.

4.5.5 THE IMPORTANCE OF ISOLATION

A large population occupying a large area **evolves slowly,** as random interbreeding tends to maintain an average type. But if, at a given moment, this population is divided into several geographically separate groups, each group will continue to evolve separately.

Isolation also creates small populations, each following its own evolutionary path, different from that of the other group. This is how **varieties** or distinct **species** are born from the initial population.

Isolation must be considered an important factor in diversification.

Indeed, for a group to invade a new environment, it must already possess structures that enable it to live in that environment. For French genetic biologist **Lucien Cuénot**, the necessary adaptation must precede settlement in a given environment, and is always a pre-aptitude, a pre-adaptation.

Lucien Cuénot
(1866-1951)

Isolated populations are subject to new environmental influences and present somations and specific mutations that are not identical in all regions. As ecological conditions are not identical, natural selection will not ensure the predominance of the same genotype.

4.5.6 MICROEVOLUTION AND MACROEVOLUTION

Variations caused by the environment can lead to the formation of lower subdivisions in the classification: race or variety, new species, we speak of **microevolution**, because the modifications are not very important.

Macroevolution is that which reveals the major types of organization corresponding to the higher groups in the classification: genera and phyla.

For the synthetic theory, the mechanism remains the same: it's the significant variations in environmental conditions that create breaks in the adaptive equilibrium that impose new modifications, which explain the appearance of new types.

4.5.7 ENDOGENOUS FACTORS

Non-chromosomal environmental factors are not enough to explain evolution. It also involves **internal factors**.

Chromosomes and **genes** are the most important factors in heredity; they are the carriers of hereditary factors, but it should be noted that certain constituents of living matter are carriers of their own heredity.

The **physico-chemical state of** the cytoplasm is also involved in gene expression, and conditions the success and outcome of synthesis. Together, **genome** and **cytoplasm**, they are necessary for the formation of an individual.

4.5.8 GENETIC RECOMBINATION

Genetic recombination causes genes to associate differently, producing different phenotypes, which are then influenced by the environment and natural selection. These can favor or eliminate one gene or another.

4.5.9 CRITICISM OF SYNTHETICAL THEORY

Synthetic theory represents a considerable advance in our understanding of evolution.

But this cannot be considered the definitive explanation of the mechanism of evolution. There are still many obscure points about the mechanism of evolution.

Until now, it has been difficult to understand how mutations and natural selection could create entirely new organs, when their actions consist in modifying and amplifying already existing functions.

The mechanism of mutation as we have seen it up to now does not explain the parallel evolution of two **coapted organs**. **Coaptations** are adjustments of two organs of different origin. Male and female copulatory organs are a good example of coaptation.

The combination of mutation and natural selection does not take the species out of its context.

Synthetic theory doesn't explain the **parallel** evolution of complex organs like the eye in different lineages, nor does it explain the many cases of **convergent** evolution or **extinction**.

In the end, an organized world cannot emerge from chance, and the determinism of macroevolution still eludes biologists.

4.6 THE NEUTRALIST THEORY OF EVOLUTION

4.6.1 PRINCIPLE

This theory postulates that genetic mutations escape natural selection and spread independently within populations. Indeed, many mutations do not affect proteins and their functions, either because they concern non-coding DNA, or because they modify an amino acid without modifying the codon involved. Consequently, they induce **neither selective advantage nor disadvantage**.

The transmission of these neutral mutations therefore takes place according to other parameters (such as chance), which become just as important as natural selection. Neutralist theory explains part of genetic diversity.

In molecular biology, the theory that most genetic polymorphism is the consequence of **neutral mutations** with no particular adaptive value; molecular evolution within populations therefore depends on random genetic drift alone.

The **neutralist theory of evolution**, also known as the **theory of mutation and random drift**, is a theory of molecular evolution according to which most mutations are neutral and have a negligible influence on selective value.

It explains genetic diversity mainly in terms of genetic drift, and gives natural selection only an occasional role, without however contesting its preponderance from the point of view of morphological evolution.

Motoo Kimura
(1924-1994)

It was formalized by **Motoo Kimura** from 1968 onwards, before becoming one of the pillars of molecular evolution. In particular, **the neutrality hypothesis**, *under which mutations have no influence on selective value, is the null hypothesis generally retained in work where such an assumption is necessary.*

Based on mathematical models of diffusion, **Kimura** postulated that certain genetic mutations lead to changes at the molecular (protein) level that are neutral with regard to natural selection. This depends on the mutation's place in the DNA nucleotide sequence. 72% of synonymous (silent) mutations involve a nucleotide in third position.

In other words, these mutations are selectively neutral, as the different versions of the encoded proteins do not change the organism's adaptability to its environment. Similarly, in eukaryotes, there are non-coding sequences called introns, and mutations occurring in these regions are also without effect or neutral.

From a methodological point of view, showing that a mutation has a different adaptive value from other mutations is tantamount to testing the hypothesis of neutrality according to diffusion models and rejecting this hypothesis.

In other words, the diffusion of the mutation within the population is calculated using the equations provided by the theory, and the result obtained is compared with observations using statistical tests. If the theoretical results differ from the observations, the mutation has an impact on the adaptability of the population under consideration.

One of the implications of this theory is that evolution does **not** necessarily **lead** to organisms becoming more complex. In the simplest organisms, such as prokaryotes, evolution leads to more complex organisms because of the **left wall,** i.e. the minimum viable limit, of complexity.

But from a certain stage of complexity onwards, this theory predicts **possible involutions**. Some spider species, for example, have lost their social behavior during evolution, which corresponds to a simplification of the interactions of these organisms with their fellow creatures.

4.6.2 NEUTRALISTS AND SELECTIONISTS

Although some have perceived this theory as anti-Darwinian, Kimura and most evolutionary biologists today agree that the two theories are **compatible**.

Indeed, in this theory, natural selection **loses** its status as the dominant evolutionary factor and becomes just one factor among others, including **stochastic factors** such as genetic drift.

The role of natural selection is not, however, called into question by neutralist theory.

4.6.3 DISCUSSION OF THE THEORY

The neutralist theory of evolution has both **strengths** and **weaknesses**

One of the **strengths of** the neutralist theory is that it encompasses the synthetic theory of evolution. What's more, it can explain a number of observed phenomena.

But the calculations used to develop the mathematical models on which this theory is based are flawed: they are based on strong assumptions, some of which are not systematically valid. Neutralist theory assumes that :

- The mutation **rate** is constant over time.
- Population **size** is constant over time.
- The mutation-drift **equilibrium** is verified: the number of alleles lost through genetic drift is equal to the number of new alleles produced by mutation.

The mutation **rate** is not constant throughout the DNA. But if we restrict ourselves to particular sequences, then we can assume that the mutation rate becomes homogeneous on average over the long term. Even if in the short term, as under certain stress conditions, for example, this rate fluctuates from time to time.

We also know that population size is generally fairly constant, even if this is not guaranteed on **geobiological** time scales such as during ecological upheavals, i.e. periods of **mass extinction** or **evolutionary radiation** when this population size is punctually disrupted.

We know that the mutation-drift equilibrium state is largely respected, since the genetic distance between the different phyla remains very small.

REVISION QUESTIONS

1. How can we sum up the major transformations that have marked the evolution of Equidae?
2. What were the first living things? How did they form? What is their approximate age?
3. What does fixism mean? transformism?
4. How can we explain the similarities between the fauna and flora of the two continents?
5. Why do evolutionists attach so much importance to Devonian crossopterygian fish?
6. Explain why the Secondary period is called the Age of Reptiles?
7. Trace the major stages in the evolution of plants and animals.
8. Explain the following phenomena: relay, orthogenesis, convergence, parallelism, extinction. Give an example of each.
9. What were the first known vertebrates? Were they fish?
10. What are the main characteristics that differentiate man from Anthropoid Primates? is there a relationship between them?
11. Two classes of vertebrates are descended from Reptiles. Which two?
12. A river crosses the plain and then enters a cave, where it's always dark. In both places, on the river, we find Fish of the same species, but those on the plain have normal eyes and green coloration, while those in the cave are devoid of eyes and pigmentation.
    What would be the explanations according to : Lamarck? Darwin?
13. Is Lamarck's theory sufficient to explain evolution? What is the viewpoint of paleontologists?
14. One of the main problems posed by the study of evolution is the interdependence of germ and soma.
    - What is Lamarck's point of view?
    - What is Darwin's point?
    - What's the neo-Darwinist point of view?
15. What criticisms can be levelled at De Vries' mutationism?
16. What is the neo-Darwinist thesis based on?
17. What are the 3 main sources of hereditary variation?
18. If we were to apply Darwin's theory to the wild lion, it would have to become stronger and stronger from one generation to the next. Discuss.
19. Between the ventral scales, the python has preserved the remains of its legs (claws, coxal bone, femur...). Explain this, according to Lamarck, according to neo-Darwinians.
20. What role does isolation play in the evolution of species? Illustrate

your answer with a concrete example.

21. At a certain stage of development, all vertebrates have gill slits, but only aquatic animals have them. What evolutionary theory can this fact support? Justify your answer.
22. The Giraffe has a very long neck, which would appear to be a consequence of evolution, because this animal has long been continually stretching to reach the succulent leaves of trees. Comment.
23. The Kiwi is a small New Zealand runner, its wings persisting only in trace form. How do evolutionists interpret these useless structures?
24. Synthetic theory cannot explain all the problems posed by evolution. Which ones, for example? can we question the very fact of evolution? explain.
25. Explain these concepts: mass extinction, evolutionary radiation, geological time, stochastic factor, left wall and neutral mutation.
26. On what principle is the neutralist theory of evolution based?
27. Beg the question of **Kimura**'s neutrality.
28. Contrast the neutralist theory with the theory of natural selection.
29. What are the strengths and weaknesses of the neutralist theory of evolution?

BIBLIOGRAPHICAL REFERENCES

BOOKS CONSULTED IN ENGLISH

A.J. Grove, G.E. Newell, J.D. Carthy, E.H. Mercer and P.F. Newell. Animal Biology. Eighth Edition, University Tutorial Press LTD London. 910p

Ann Fullick, Halima Mwinshekhe, Mary Muddu, Seline Nyabua. *A Level Biology,* Longman Part of Pearson ;2011. 488p

D.G. Mackean. *Introduction to Biology,* Third Tropical Edition, John Murray London; 2006. 256p

Dictionary of Contemporary English, New Edition for Advanced Learners. Longman/Person, Edinburgh; 2010. 2082p

Kenneth R. Miller and Joseph Levine. *Biology,* Pearson Prentice Hall, Boston ;2006. 1146p

Lee Ching and J. Arunasalam. *Biology,* Volume 2. Pre-U text STPM, Pearson Longman ;2008.408p

MBV Roberts, *Biology a Functional Approach,* 4th Edition, Nelson CHELTENHAM ;1986. 693p

Michael Kent. *Advanced Biology,* 2nd Edition, Oxford University Press; 2013. 632p

Michael Roberts, Michael Reiss and Grace Monger. *Biology Principles and Processes,* Nelson Edinburgh ;1993. 852p

MVB Roberts and TJ King. *Biology: A Functional Approach.* Students' Manual, 2nd Edition. Nelson1987. 416p

Neil A. Campbell and James B. Reece. *Biology,* Eighth Edition, Pearson/Benjamin Cummings San Francisco2008. 1267p with Glossary and Index.

Pete Kennedy and Frank Sochacki. *Biology,* OCR Heinemann, Edinburgh; 2008. 273p

Peter H. Raven and George B. Johnson. *Biology,* Sixth Edition. Mac Graw Hill, Boston ;202. 1238p with Appendix, Glossary and Index.

CONSULETS IN FRENCH

COMBALUZIE C. *Introduction à la géologie,* Ed. Au Seuil, Paris 1961. 192 pages.

PETIT C. and PREVOT G. *Génétique et évolution,* Herman, Paris 1970. 392 pages GEVAERESTS H. *Biologie générale,* Université de

Kisangani, 1985. 334 pages DEYSSON G. *Cours de botanique générale,* SEES, Paris 1964. 336 pages CAMEFORT H. et GAMA A., *Sciences naturelles,* Hachette, Paris 1953. 670 pages

LEHMAN J.P., *Les preuves paléontologiques de l'évolution,* PUF, Paris 1970. 476 pages

LIMOGES C. La *sélection naturelle,* PUF, Paris, 1970. 184 pages

DESIRE Ch. et al. *Sciences naturelles,* Bordas, Paris 1974. 488 pages

MAGLOIR G. *Theillard et le Sinanthrope* Ed. Universités, Paris 1964. 64 pages OPARINE A., *L'origine de la vie sur terre,* Masson, Paris 1965

FLORI J. et al. *Evolution or creation?* Ed. SDT Lys 1974. 381 pages

Wilhelm Nultsch, *Botanique Générale,* 10th Edition, De Boeck Université, Thème Verlag; 1998. 602p

André Beaumont and Pierre Cassier. *Travaux pratiques de Biologie Animale,* Zoologie Embryologie Histologie. Dunod Paris ;1998 502p

Bernard and Geneviève Pierre. *Dictionnaire Médicale pour les Régions Tropicales.* Bureau d'Etudes et de Recherches pour la Promotion de la Santé, Kangu-Mayumbe/Republic of Zaire; 1987.871p

Cain, Damman, Lue, Yoon. *Découvrir la Biologie*, 2nd Edition, De Boeck Bruxelles; 2006. 728p with glossary and index.

Max Aron and Pierre-Paul Grassé. *Biologie Animale*, Masson et Cie Editeurs, Pairs 1966.1421p

Paul Duvignaud. *La Synthèse Ecologique*: Population, Communauté, Biosphère, Noosphère, Doin Paris; 174. 296p.

MENANT G.F. et G. *Ecologie Pays intertropicaux,* 1ère A B D. Hatier ; 1975 Paris. 80 p

DAJOZ R. *Précis d'Ecologie*, Dunod Paris ; 1970. 358p

Jean-Marie Pelt. *Le Tour du Monde d'un Ecologiste.* Fayard,1990. 488p

Le Petit Robert de la langue française. Electronic version.

The **Bible**, *New World Translation.* Revised 2018 edition of Watch Tower Bible and Tract Society of Pennsylvania. 1834 p and Appendices

Albert ALOMBA A.

Professor of chemistry and life and earth sciences at the Groupe Scolaire Consulaire Congolais in Kigali.

Telephone numbers: +250788540257 and +250738540257

E-mails: alomba2@gmail.com and alomba2@yahoo.fr

More Books!

yes

I want morebooks!

Buy your books fast and straightforward online - at one of world's fastest growing online book stores! Environmentally sound due to Print-on-Demand technologies.

Buy your books online at
www.morebooks.shop

Kaufen Sie Ihre Bücher schnell und unkompliziert online – auf einer der am schnellsten wachsenden Buchhandelsplattformen weltweit! Dank Print-On-Demand umwelt- und ressourcenschonend produzi ert.

Bücher schneller online kaufen
www.morebooks.shop

info@omniscriptum.com
www.omniscriptum.com

OMNIScriptum

MIX
Papier aus verantwortungsvollen Quellen
Paper from responsible sources
FSC® C105338

Printed by Books on Demand GmbH, Norderstedt / Germany